Lecture Notes of the Institute for Computer Sciences, Social Informatics and Telecommunications Engineering 591

Editorial Board Members

Ozgur Akan, *Middle East Technical University, Ankara, Türkiye*
Paolo Bellavista, *University of Bologna, Bologna, Italy*
Jiannong Cao, *Hong Kong Polytechnic University, Hong Kong, Hong Kong*
Geoffrey Coulson, *Lancaster University, Lancaster, UK*
Falko Dressler, *University of Erlangen, Erlangen, Germany*
Domenico Ferrari, *Università Cattolica Piacenza, Piacenza, Italy*
Mario Gerla, *UCLA, Los Angeles, USA*
Hisashi Kobayashi, *Princeton University, Princeton, USA*
Sergio Palazzo, *University of Catania, Catania, Italy*
Sartaj Sahni, *University of Florida, Gainesville, USA*
Xuemin Shen, *University of Waterloo, Waterloo, Canada*
Mircea Stan, *University of Virginia, Charlottesville, USA*
Xiaohua Jia, *City University of Hong Kong, Kowloon, Hong Kong*
Albert Y. Zomaya, *University of Sydney, Sydney, Australia*

The LNICST series publishes ICST's conferences, symposia and workshops.
LNICST reports state-of-the-art results in areas related to the scope of the Institute.
The type of material published includes

- Proceedings (published in time for the respective event)
- Other edited monographs (such as project reports or invited volumes)

LNICST topics span the following areas:

- General Computer Science
- E-Economy
- E-Medicine
- Knowledge Management
- Multimedia
- Operations, Management and Policy
- Social Informatics
- Systems

Marta Ziosi · Giovanni Sartor ·
João Miguel Cunha · Angelo Trotta ·
Philipp Wicke
Editors

AI for People, Democratizing AI

Second EAI International Conference, CAIP 2023
Bologna, Italy, November 24–26, 2023
Proceedings

Editors
Marta Ziosi
University of Oxford
Oxford, UK

Giovanni Sartor
University of Bologna
Bologna, Italy

João Miguel Cunha
University of Coimbra
Coimbra, Portugal

Angelo Trotta
University of Bologna
Bologna, Italy

Philipp Wicke
Ludwig Maximilian University of Munich
Munich, Germany

ISSN 1867-8211 ISSN 1867-822X (electronic)
Lecture Notes of the Institute for Computer Sciences, Social Informatics
and Telecommunications Engineering
ISBN 978-3-031-71303-3 ISBN 978-3-031-71304-0 (eBook)
https://doi.org/10.1007/978-3-031-71304-0

© ICST Institute for Computer Sciences, Social Informatics and Telecommunications Engineering 2024

This work is subject to copyright. All rights are solely and exclusively licensed by the Publisher, whether the whole or part of the material is concerned, specifically the rights of translation, reprinting, reuse of illustrations, recitation, broadcasting, reproduction on microfilms or in any other physical way, and transmission or information storage and retrieval, electronic adaptation, computer software, or by similar or dissimilar methodology now known or hereafter developed.
The use of general descriptive names, registered names, trademarks, service marks, etc. in this publication does not imply, even in the absence of a specific statement, that such names are exempt from the relevant protective laws and regulations and therefore free for general use.
The publisher, the authors and the editors are safe to assume that the advice and information in this book are believed to be true and accurate at the date of publication. Neither the publisher nor the authors or the editors give a warranty, expressed or implied, with respect to the material contained herein or for any errors or omissions that may have been made. The publisher remains neutral with regard to jurisdictional claims in published maps and institutional affiliations.

This Springer imprint is published by the registered company Springer Nature Switzerland AG
The registered company address is: Gewerbestrasse 11, 6330 Cham, Switzerland

If disposing of this product, please recycle the paper.

Preface

We are delighted to introduce the proceedings of the second edition of the European Alliance for Innovation (EAI) International Conference on AI for People (CAIP). CAIP 2023 was co-organized by the non-profit association AI for People and by the European Alliance for Innovation (EAI) and held at Palazzo Dal Monte, University of Bologna, Bologna (Italy), 24–26 November 2023.

In this year's edition, the conference had the theme Democratizing AI, which refers to the democratization of its stages of design, development, deployment and use. The conference provided its participants with opportunities to gain a better understanding of and critically examine what 'democratizing AI' entails.

For the scientific program, CAIP 2023 received 27 submissions, of which 8 papers were accepted and orally presented at the conference. In addition to the main track, a secondary work-in-progress track was created and 5 submissions were accepted for presentation at the conference. The paper presentations were divided into 4 thematic sessions, aligned with the main theme of the conference (Democratizing AI): Session 1 on Data Privacy and Technology Ethics; Session 2 on Democratization of AI and Governance; Session 3 on AI in Society and Legal Aspects; and Session 4 on Ethical AI and Innovation.

Although multiple CAIP 2023 submissions were of high quality, the organizers decided to recognize the quality of one of the papers with a Best Paper award, given to the paper "(Im)possibilities in the Ethics of AI: Biometric Surveillance, Complicity, and Refusal in India and Beyond" by Nikhil Dharmaraj.

In addition to the paper presentation sessions, the program also featured two keynote speeches and three invited talks. On November 24, the first keynote speaker, Sabrina Küspert (Mercator Foundation), delivered a talk titled "How to govern AI? Risks and Responsibilities along the Value Chain". The second keynote speaker was Elizabeth Seger (Centre for the Governance of AI; Centre for the Study of Existential Risk, University of Cambridge), who spoke on November 25 on the topic "AI Democratization: more than Model Dissemination".

For the three invited speakers, the conference organizers decided to invite local associations which could share their perspectives at the intersection of AI and society. Associations included Tech Workers Coalition Italia (twc-italia.org), Privacy Network (privacy-network.it), and Into the Black Box (http://www.intotheblackbox.com/). On November 25, the program included a session titled "Getting to know AI for People" in which the Association AI for People was introduced to participants, who then joined a Q&A session.

CAIP 2023 also had part of its program dedicated to social events. This year, two specific social events took place: a walking tour of the city of Bologna, which showed participants some of the iconic buildings; and a conference dinner at MamBo (the Modern Art Museum of Bologna) with a performance by the Band Divinae Miranda.

Acknowledgments

The CAIP 2023 organizational team expresses our gratitude to our sponsors:

- University of Bologna, Department of Computer Science and Engineering (DISI)
- EU Project TAILOR (Foundations of Trustworthy AI–Integrating Reasoning, Learning and Optimization)

We are particularly grateful to the CIRSFID (University of Bologna), which offered organizational assistance to the conference and made available the rooms at Palazzo Dal Monte where the conference took place.

March 2024

Marta Ziosi
Giovanni Sartor
João Miguel Cunha
Angelo Trotta
Philipp Wicke

Organization

Steering Committee

Marta Ziosi	University of Oxford, UK
João Miguel Cunha	University of Coimbra, Portugal

Organizing Committee

General Chair

Marta Ziosi — University of Oxford, UK

General Co-chair

Giovanni Sartor — University of Bologna, Italy

TPC Chair and Co-chairs

Angelo Trotta	University of Bologna, Italy
João Miguel Cunha	University of Coimbra, Portugal
Philipp Wicke	Center for Information and Language Processing (CIS), Germany

Sponsorship and Exhibit Chair

Angelo Trotta — University of Bologna, Italy

Local Chair

Gabriele Graffieti — University of Bologna, Italy

Publicity and Social Media Chair

João Miguel Cunha — University of Coimbra, Portugal

Publications Chair

Angelo Trotta University of Bologna, Italy

Web Chair

João Miguel Cunha University of Coimbra, Portugal

Contents

Ethical AI and Innovation

A Human-Centered Decision Support System in Customer Support 3
   Sven Münker, Marcos Padrón, Antonia Markus, Marco Kemmerling,
   Anas Abdelrazeq, and Robert H. Schmitt

An AI-Based Remote Rehabilitation System to Promote Access to Physical
Rehabilitation 11
   C. Gmez-Portes, S. Martínez, S. Schez-Sobrino, V. Herrera,
   J. A. Albusac, and D. Vallejo

Democratization of AI and Governance

Democratization is a Process, not a Destination: Operationalizing Ethics
and Democratization in a Cyberinfrastructure for AI Project 29
   Sadia Khan, Alfonso Morales, and Beth Plale

(Im)possibilities in the Ethics of AI: Biometric Surveillance, Complicity,
and Abolitionist Refusal in India and Beyond 46
   Mallika G. Dharmaraj

Research Methods of the Impact of AI on Elections – Systematic Review 63
   Maria Lipińska

AI in Society and Legal Aspects

On the Legal Aspects of Responsible AI: Adaptive Change, Human
Oversight, and Societal Outcomes 73
   Daria Onitiu, Vahid Yazdanpanah, Adriane Chapman, Enrico Gerding,
   Stuart E. Mid-dleton, and Jennifer Williams

Limitations of Transparency in Democratising and Regulating Algorithmic
Management 86
   Miranda Cross

Breaking the Filtered Lens: A Feminist Examination of Beauty Ideals
in Augmented Reality Filters 95
   Mariana P. Castillo-Hermosilla, Hedye Tayebi-Jazayeri,
   and Victoria N. Williams

Data Privacy and Technology Ethics

Your Body Should Not Belong to the Internet: Online Bodily Integrity
in the World of Deepfake Pornography 105
   Lyndsey Scott

The Promise and Peril of Responsible AI Principles 117
   Retno Larasati

Accelerating Machine Learning Primitives on Commodity Hardware 125
   Roman Snytsar

Author Index 131

Ethical AI and Innovation

A Human-Centered Decision Support System in Customer Support

Sven Münker(✉), Marcos Padrón, Antonia Markus,
Marco Kemmerling, Anas Abdelrazeq, and Robert H. Schmitt

RWTH Aachen University, Aachen, Germany
{sven.muenker,m.padron,a.markus,marco.kemmerling,
anas.abdelrazeq}@ima.rwth-aachen.de, r.schmitt@wzl-mq.rwth-aachen.de

Abstract. In customer support, employees have to handle a high number of customer e-mail inquiries every day. With an increasing variety of products, the number of different possible customer problems increases significantly. The employees face a sheer unmanagable task to distinguish between the different customer problems. In order to improve customer satisfaction and reduce employee stress and workload, we present a human-centered and artificial intelligence-based decision support system for employees in customer support. The system can be deployed towards classifying customer e-mail inquiries into pre-defined categories. For this purpose, we utilize a natural language processing pipeline around a recurrent neural network returning to the employees the best-fitting categories for incoming e-mails.

Keywords: Human-Centered AI · Natural Language Processing · Decision Support System · Customer Service

1 Introduction

Customer service agents are expected to answer quickly. Due to an increasing variety of products and in this context an exploding number of associated possible problems, customer service agents face increasing workloads, which results in heightened stress levels and compromised efficiency.

Recent advancements in Natural Language Processing (NLP), Machine Learning (ML) and computational power present a promising solution to alleviate these challenges. NLP and ML have been used for automated customer service solutions like chatbots [1] and automated e-mail answering systems [2], or for detecting unusual mails [3]. Due to the complexity and variety of customer problems in technical customer support, automated systems often do not

---

The research presented in this paper has been carried out within the research project "AIXPERIMENTATIONLAB" (Project number EXP.01.00016.20). The authors gratefully acknowledge the support of the German Federal Ministry of Labor and Social Affairs (BMAS) and the Initiative New Quality of Work (INQA).

have the required accuracy and cannot give individual answers to the customers. Therefore, decision support can be leveraged to focus on assisting the employees in their workflow instead of automating the workflow. Decision support systems (DSS) aim to support the users in complex tasks. In customer support, support systems have been developed to train novice customer agents [4], to sort incoming e-mails to customer support agents in correspondence to their technical expertise [5] or by predicting the most likely answer to a question [6].

In this paper, we present a DSS that uses NLP and ML technologies and is intended to reduce the stress and workload experienced by customer service agents. Based on a human-centered approach, we aim to reduce the search time for possible answers by proposing solutions from pre-defined problem categories. We process the incoming e-mails through an NLP pipeline that embeds a recurrent neural network classifier and returns the top-3 best candidates to the users. The use of the system is optional, so it is not forced into the employee's workflow and the employee is free to use or discard the system's suggestion.

The paper is strucured as follows: First, we present the current workflow of the customer service employees and the associated challenges to be solved by the DSS. Next, we present our development approach for a DSS along with the associated dataset. Finally, we present the employees' workflow including the DSS.

2 Development Approach for a Decision Support System in Customer Support Setting

Our DSS was developed in close cooperation with a German medium-sized company. To understand the real-life challenges encountered by customer service employees, we followed the human-centered approach (as described in DIN EN ISO 9241-210). First, we evaluated the employees tasks on-site in the company throughout the workday. These workshops were performed over multiple days, to better capture the employees' working environment. During the visit, we used the Day-In-Life-Of method, in which we observe the employees' tasks in their workspace without interacting with them, to get a realistic view on their challenges. Additionally, we used cognitive walkthroughs [7], where the employees explained every step of their critical tasks in detail. Afterwards, we extracted the first prototype of the use-case for the decision-support system. This use-case was adapted during the technical development process through frequent feedback iterations to see if the use case matched the workflow of the employees.

2.1 Pre-system Workflow

The initial workflow of the employees before introducing the decision-support system (see Fig. 1) is as follows: The employee gets an e-mail with a customer inquiry. If the employee knows the answer to the customer's question, they answer directly. Otherwise they look up the categories in their frequently asked questions (FAQ) catalogue. The catalogue consists of frequent customer

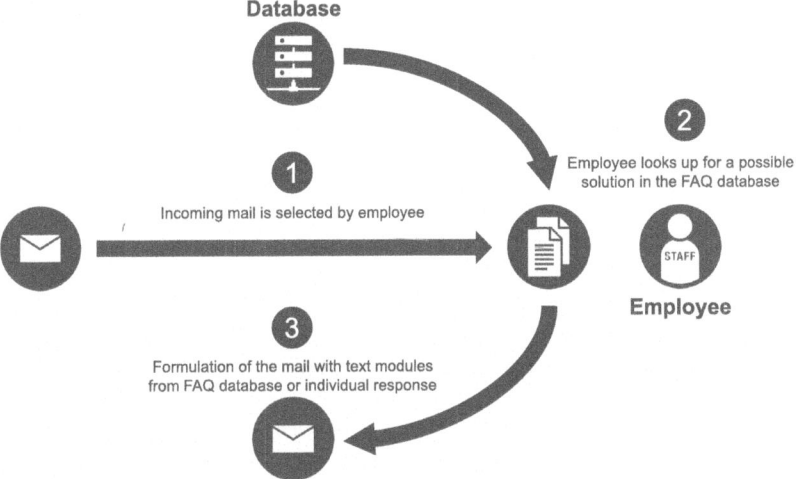

Fig. 1. Customer service agent workflow before the implementation of the decision support system

questions and pre-formulated answer prototypes. The employee can copy the pre-formulated answer, adapt it to fit the customer's demands and answer the customer. If the employee cannot find a fitting solution, they can either ask the user for more information, or escalate it to the second support level.

After the e-mail is answered by the employee, it is saved in the companies database and labeled with the corresponding category number. If the e-mail does not fit to a pre-existing category, a new category is created. If multiple categories were used to answer the e-mail, one copy of the e-mail is saved for every category.

The employees face multiple challenges in their workflow: On the one hand, the customers expect quick answers to their inquiries. On the other hand, the customers' e-mails range from short e-mails without the necessary details to long e-mails including a lot of details that are irrelevant to the problem. This is represented by the large number of over a 1000 categories in the FAQ-database, of which 471 are used actively.

Before the implementation of the DSS, we performed a pre-measurement of the employees stress level using a questionaire [8].

2.2 Decision Support System

To support the employees decision process and to help them make better and faster use of the FAQ-database, we designed a DSS. The system is supposed to reduce the time of searching the database for the corresponding category, by pre-clàssifying the incoming e-mails into the FAQ-categories. Afterwards, the employees is still required to modify the pre-formulated answer of the chosen category, to best fit the customer's needs. This step is obligatory, due to complex technical inquiries and to ensure high quality in customer support. It is important to note, that the system is optional, such that it is up to the employee to decide: First, whether they want to get system suggestions for a specific e-mail and second, if they want to follow the given suggestion.

2.3 Technical Solution

To build the backend of the customer support system, we use an ML based classification model, which is implemented within a NLP pipeline. The model is trained on an FAQ-labeled dataset of customer e-mails. This dataset was built in the last two years by employees in the way described in Sect. 2.1. In the following, we present details of the dataset and illustrate the NLP pipeline.

The e-mail dataset consists of 117321 customer e-mail conversation. Some conversations are sorted into multiple categories, in that case a copy for each category is saved. There are 471 different categories that range from general organisational questions to highly product specific technical problems. The dataset is highly imbalanced, with the majority of the e-mails (54%) belonging to the 10 highest categories and 188 categories having 10 or less entries. The e-mails varies between zero and 65320 characters, with an average of 2297 characters per mail. The word-count per e-mail is between zero and 8540 words, with an average of 269 words per mail. The e-mails have between 1 and 6 categories, with an average of 1.2 and a median of 1.

Our NLP pipeline consists of six consecutive steps:

1. **Removing multi-labeled e-mails**: We analyze how many categories there are in each e-mail. While most of the e-mails (91%) only have a single label, some e-mails belong to multiple categories. Due to the high sparsity of multi-labeled mails, we decided to follow a single-class classification approach. Therefore we removed all e-mails from the dataset that had multiple labels and kept only those, where the included information was deemed sufficient to categorize it into a single category. After removing the multi-labeled e-mails, we remained 88610 e-mails.
2. **E-mail splitting**: To avoid whole e-mail conversations including employees requests and incomplete answers, we split such e-emails and discard most of the e-mail history despite the last customer e-mail. In particular, before e-mail splitting the dataset had on average 2.0 e-mails per datapoint with a minimum of one and a maximum of 51 e-mails. To split the dataset entries into seperate e-mails, we identify the beginning and the end of each subsequent e-mail by their greeting and farewells.
3. **Pre-processing**: To reduce variability in the model input, we perform a pre-processing based on the NLP packages NLTK [9] and Textacy [10]. We replace mutuations in the text, remove encoding errors and use lowercasing to unify the text. We also remove punctuation, numbers, headers, e-mail addresses and websites. To further reduce the variety of words, we use lemmatization to break down the words into their infinite stem.
4. **Feature selection**: To extract features for the machine learning model, we used two different methods:

The **Count Vectorizer** (CV) is a form of the "bag of words representation" for texts. It creates a token for every word in the text string and counts how often the word is used in the string. It creates a sparse matrix with the word-tokens as columns and the different word counts as entries.

The **Byte-Pair Encoder** (BPE) [12] optimizes the encoding process of strings into token vectors. Based on a total number of words (in our case 20000), the BPE sorts character combinations into tokens. Starting from single character, over shorter words, up to longer words, depending on the total allowed vocabulary of the tokenizer. This method has the advantage, that it can not only tokenize known words from the dataset it was trained on, but also split unknown words into parts, that can still be further evaluated. This is especially useful, since the dataset is based on the German language, where compound words are very common.

5. **Training of the classification model**: For the classification, we compared 9 different machine learning models on a reduced version of the dataset. Starting from statistical models based on Bayesian classification [13] and Linear Regression [14], over more complex mathematical modeling with a Support Vector Machine (SVM) [15] and a KNN-Classifier (KNN) [16], up to deep learning models (MLP [17], RNN-BiLSTM [18]).

   To train the models, we use supervised learning. We split the dataset into different training and test sets with 90% learning data und 10% test data. We evaluated the model performance based on accuracy, precision, recall and f1-score.

6. **Post-processing**: To obtain a result that is interpretable by the employees, we postprocess the model output. Since the system is not supposed to work autonomously, but rather give suggestions to the users, we performed participatory workshops [19] to evaluate how the employees wanted the system to present the suggestions. Consequently, we give the three most promising matches of the classification algorithm as results for the employees in combination with a visual indicator or the corresponding probability.

Table 1. Performance of different system prototypes

Model	Precision	Accuracy	F-1 Score	Top-3 Score
Naive Bayes	29%	31%	28%	60%
SVM	25%	17%	15%	51%
Lin. Regression	32%	33%	29%	64%
RNN-BiLSTM	69%	70%	69%	89%
MLP	50%	49%	48%	72%

A comparison of the model performance can be found in Table 1. Based on the combination of performance and low runtime, we chose the RNN classifier in combination with a byte-pair encoding feature extractor for the final system.

The RNN has a precision of 68% and a Top-3 score of 87%, which is comparable to other research [5,20]. The python backend is attached to the companies server and returns the suggestions in form of the top-3 best candidates for the input mail.

2.4 Post-system Workflow

To establish the decision-support system in the company, we developed a frontend together with the employees in participatory workshops, to best fit the users' needs. The new workflow including the decision-support system is presented in Fig. 2. After getting an e-mail the employee can still follow the known workflow from Sect. 2.1. If they want to use the decision-support system, they can click on a button and get suggestions for the best-fitting FAQ-categories based on our trained backend. Since the system is not error free and thus sometimes makes wrong suggestions, the employee is free to use or discard the suggestions and follow the steps from the previous workflow (see Sect. 2.1). Besides suggestions for the associated FAQ number, the frontend also gives a pre-formulated answer prototype, which then needs to be adjusted by the employee.

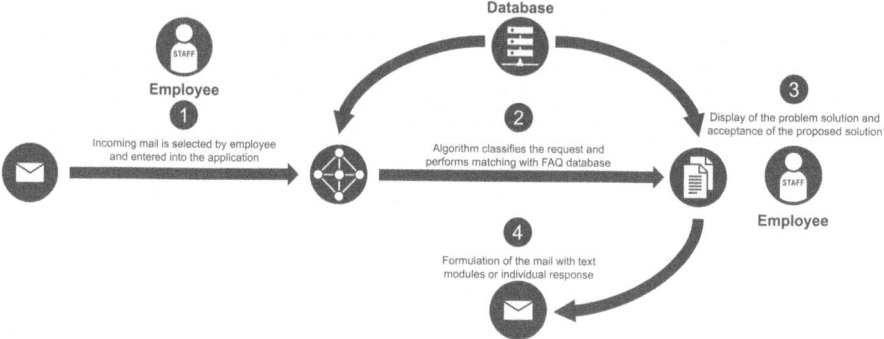

Fig. 2. Customer Service Agent Workflow after the implementation of the decision support system

3 Conclusion

In this short paper, we presented a human-centered DSS for customer support agents. The system is intended to reduce stress and workload of the customer support agents, while maintaining their usual workflow by giving them the choice to use the system. The backend consists of an NLP pipeline which embedds an RNN classifier. Our pipeline and classifier are shown to perform comparable to other studies.

The system is under evaluation in the company and a post-measurement of the employees' stress levels will be performed after eight weeks. It will be compared with the pre-measurement to evaluate the system's influence on the users' stress levels.

In future work, we aim to improve our backend performance by rebalancing techniques for the dataset, to get higher accuracy results for FAQ-categories with a low number of entries in the database. This can be achieved by creating new e-mails from underrepresented categories using generative AI models, or using a multi-label classifier that proposes suggestions for the different labels, without excluding other categories. A second research direction could be to frequently update the database and use online learning to adapt to new categories.

References

1. Cui, L., Huang, S., Wei, F., Tan, C., Duan, C., Zhou, M.: SuperAgent: a customer service chatbot for e-commerce websites. In: Proceedings of ACL 2017, system demonstrations, pp. 97–102 (2017)
2. Sneiders, E.: Automated email answering by text pattern matching. In: Advances in Natural Language Processing: 7th International Conference on NLP, IceTAL 2010, Reykjavik, Iceland, August 16–18, 2010 7, pp. 381–392. Springer Berlin Heidelberg (2010). https://doi.org/10.1007/978-3-642-14770-8_41
3. Borg, A., Ahlstrand, J.: Detecting non-routine customer support E-Mails. In: 23rd International Conference on Enterprise Information Systems (ICEIS), Virtual, Online, APR 26–28, 2021 (No. 23rd International Conference on Enterprise Information Systems (ICEIS), pp. 387–394). SciTePress (2021)
4. Reinhard, P., Wischer, D., Verlande, L., Neis, N., Li, M.: Towards designing an AI-based conversational agent for on-the-job training of customer support novices. In: International Conference on Design Science Research (DESRIST). Pretoria, South Africa (2023)
5. Zicari, P., Folino, G., Guarascio, M., Pontieri, L.: Discovering accurate deep learning based predictive models for automatic customer support ticket classification. In: Proceedings of the 36th Annual ACM Symposium on Applied Computing, pp. 1098–1101 (2021)
6. Scheffer, T.: Email answering assistance by semi-supervised text classification. Intell. Data Anal. **8**(5), 481–493 (2004)
7. Mahatody, T., Sagar, M., Kolski, C.: State of the art on the cognitive walkthrough method, its variants and evolutions. Int. J. Hum. Comput. Interact. **26**(8), 741–785 (2010)
8. Buschmeyer, K., Hatfield, S., Zenner, J.: Psychological assessment of AI-based decision support systems: tool development and expected benefits. Frontiers Artif. Intell. 6, 1249322 (2023)
9. Bird, S.: NLTK: the natural language toolkit. In: Proceedings of the COLING/ACL 2006 Interactive Presentation Sessions, pp.69–72 (2006)
10. PYPI Homepage. https://pypi.org/project/textacy/. Accessed 13 July 2023
11. SciKit-Learn Homepage. https://scikit-learn.org/stable/modules/generated/sklearn.feature_extraction.text.CountVectorizer.html#sklearn.feature_extraction.text.CountVectorizer. Accessed 13 July 2023
12. Shibata, Y., et al.: Byte Pair encoding: a text compression scheme that accelerates pattern matching(1999)

13. Rennie, J.D.M.: Improving multi-class text classification with naive Bayes (2001)
14. Zhang, T., Oles, F.J.: Text categorization based on regularized linear classification methods. Inf. Retrieval **4**, 5–31 (2001)
15. Joachims, T.: Text categorization with support vector machines: learning with many relevant features. Eur. conf. mach. learn. Berlin, Heidelberg: Springer Berlin Heidelberg (1998). https://doi.org/10.1007/BFb0026683
16. Yong, Z., Youwen, L., Shixiong, X.: An improved KNN text classification algorithm based on clustering. J. Comput. **4**(3), 230–237 (2009)
17. Rosenblatt, F.: Principles of neurodynamics. perceptrons and the theory of brain mechanisms. Cornell Aeronautical Lab Inc Buffalo NY. (1961)
18. Xu, G., Meng, Y., Qiu, X., Yu, Z., Wu, X.: Sentiment analysis of comment texts based on BiLSTM. IEEE Access **7**, 51522–51532 (2019)
19. Spinuzzi, C.: The methodology of participatory design. Tech. Commun. **52**(2), 163–174 (2005)
20. Van Landeghem, J., Blaschko, M., Anckaert, B., Moens, M.F.: Benchmarking scalable predictive uncertainty in text classification. IEEE Access **10**, 43703–43737 (2022)

An AI-Based Remote Rehabilitation System to Promote Access to Physical Rehabilitation

C. Gmez-Portes, S. Martínez, S. Schez-Sobrino, V. Herrera, J. A. Albusac, and D. Vallejo

Department of Computer Science, University of Castilla-La Mancha,
Paseo de la Universidad 4, 13071 Ciudad Real, Spain
David.vallejo@uclm.es

Abstract. Remote physical rehabilitation for patients affected by neurological diseases is a promising research line, contributing to improving healthcare delivery, addressing the shortage of qualified staff, and enhancing the quality of life for patients. This paper presents a proposal for a remote rehabilitation system based on the use of Artificial Intelligence and Augmented Reality, designed to be highly accessible and scalable. The system comprises two applications: the patient application, which allows the recognition and evaluation of physical rehabilitation exercises, and the therapist application, that enables the assignment of personalised rehabilitation routines and the generation of automatic recommendations to adapt these routines according to the patient's progress level. The design aspects of the architecture are discussed, considering those that impact the scalability of the proposal when making it accessible to patients requiring remote physical rehabilitation.

Keywords: Remote Rehabilitation · Augmented Reality · Explainable AI · Soft Computing · Health

1 Introduction

In healthcare, the financial implications of long-term care and physical rehabilitation for patients with neurological diseases can impose significant economic stress on health systems [1]. This is particularly evident in contexts where these services are primarily funded or subsidised by the state. The direct expenses of treatment, including therapy sessions and medication, are indeed substantial, but the indirect costs amplify this burden exponentially. Indirect costs often include elements such as long-term care, which can span months or even years and require substantial resources and manpower. The high price of these indirect costs can often lead to prohibitive financial barriers, which can cause inadequate care or treatment delays for patients in need.

There is also a pressing issue regarding the lack of qualified personnel to oversee the rehabilitation process of these patients in person. This shortage translates into an increased workload for existing therapists, potentially diminishing the quality of care and increasing the risk of oversight or errors in the rehabilitation process [2]. Furthermore, the scarcity of qualified personnel can limit the availability of rehabilitation services, especially in rural or remote areas where healthcare resources are already sparse.

These problems are accentuated in low and middle income countries, where access to face-to-face physical rehabilitation therapies for patients with neurological diseases, such as stroke, is particularly limited [3]. These limitations can stem from a variety of factors, such as inadequate healthcare infrastructure, underfunding, long distances to healthcare centres, and lack of qualified personnel.

However, despite these challenges, technology presents a glimmer of optimism. Specifically, low-cost technological solutions for remote rehabilitation can significantly improve patient access to rehabilitation exercises and their efficacy. These solutions allow patients to conduct their rehabilitation exercises from home while remotely supervised by therapists.

Crucially, technological solutions that harness Artificial Intelligence (AI) to automatically evaluate the performance of physical exercises offer a significant breakthrough. AI can be used to personalise rehabilitation programmes, monitor patients' progress, provide real-time feedback, and alert healthcare professionals if problems are detected. Additionally, AI has the potential to democratise access to rehabilitation, as AI-based applications can be more affordable and widely available compared to conventional therapy methods.

However, implementing AI in physical rehabilitation also presents challenges. These include technical issues, such as the accuracy and reliability of AI systems; data privacy and security concerns; and issues of adoption and acceptance issues among patients and healthcare professionals. Moreover, a co-creative approach, involving patients, therapists, and AI developers, is essential to effectively deploy these solutions, ensuring that all parties are not only properly trained and equipped to use these technologies, but also actively participate in shaping and refining these tools to better meet therapeutic needs.

Thus, exploring low-cost, AI-based technological solutions for remote rehabilitation can be a promising research direction, contributing to improving healthcare delivery, addressing the shortage of qualified staff, and enhancing the quality of life for patients affected by neurological diseases. The potential of these solutions to democratise access to rehabilitation therapy is an exciting prospect, especially for low- and middle-income countries, but it must be approached with an awareness of the various challenges that lie ahead.

This paper presents a remote rehabilitation system based on the use of Artificial Intelligence and Augmented Reality, designed to be highly accessible and scalable. The system is capable of automatically assessing whether the patient performs the physical rehabilitation exercises previously assigned by the therapist at home. In this sense, the user only needs a laptop with a standard webcam connected to the Internet. Thanks to the underlying software, the system is able to track the patient's skeleton without the need for specific hardware. In addition, the system can automatically recommend modifications to rehabilitation routines assigned to patients and justify why the recommendation was made. This characteristic is related to the notion of Explainable AI.

The structure of the rest of this document is organised as follows. Section 2 provides an overview of the related work in the field, paying special attention to remote rehabilitation and Explainable AI. Section 3 presents the architecture of our proposed system, with particular focus on body tracking (Sect. 3.2), the designed data architecture (Sect. 3.3), and the automatic recommendations module (Sect. 3.4). Section 4 describes the most significant aspects of the patient application that is integrated into the proposed architecture. Finally, Sect. 5 concludes the paper and outlines future research lines.

2 Related Work

2.1 Remote Rehabilitation

In their study on remote rehabilitation systems, González-González et al. [4] introduced an innovative approach centred on designing an intelligent exergames-based rehabilitation system. This system is composed of two parts: an exergame player and a tool used to design the exergames. A key component of this system is a recommendation module, which scrutinises user interactions, physical history, and preferences to determine the suitable exergames for the user to perform. The module manages varying levels of difficulty and assesses user skills. The recommendation algorithm is grounded in three fundamental principles related to the patient's most recent performance (based on the last exergame performed): (i) if the performance was subpar, an easier exergame is selected; (ii) if the performance was satisfactory, a more challenging exergame is chosen; (iii) in other cases, a medium difficulty exergame is assigned. The system's evaluation, involving domain experts, users, and therapists, yielded positive outcomes in areas such as gesture-based interaction and medical applications.

In research more closely related to expert knowledge management, a tele-rehabilitation system is proposed in [5], focusing on the remote selection, evaluation, and administration of physical therapies. This work's central contribution is the development of a comprehensive tele-rehabilitation system. However, the authors notably stress the extraction and application of knowledge through an intricately defined ontology composed of 2300 classes and 100 properties, used to appropriately select exercises for each patient. This selection process utilises a knowledge base containing information about the patient's medical history and previously assigned treatment.

Alternatively, Karime et al. [6] introduced a web-based framework for wrist rehabilitation, utilising fuzzy logic to provide adaptive tasks to the patient while allowing concurrent therapist supervision. This study includes an effectiveness assessment of the framework, taking into account the adjustment of various parameters involved in the rehabilitation process within a setup that merges patient performance-based personalization of rehabilitation and therapist feedback. The application of fuzzy logic also emerges in research focused on patient rehabilitation employing robots or exoskeletons, such as the work explored in [7].

Here, a deterministic adaptive robust control-based system is introduced, with its control parameters optimised through a novel approach grounded in cooperative game theory. Fuzzy logic manages potential time-varying external disturbances, capitalising on its capability to handle uncertainty.

State machines constitute a methodology applied to automate the modification of rehabilitation exercise difficulty. Although not classified as Artificial Intelligence (AI), state machines present a straightforward strategy for regulating physical rehabilitation routines. In their research on upper extremity rehabilitation, Pinto et al. [8] employed state machines to introduce dynamic difficulty adjustment within their system. This system was composed of a series of exergames, each featuring a distinct number of preset difficulty levels. The state machine's role was to decide whether to increase, decrease, or maintain the level of difficulty. The parameters employed to gauge patient performance and alter the exergame's difficulty level were dependent on the specific exergame in play.

Capecci et al. [9] applied a Hidden Semi-Markov Model (HSMM) to evaluate rehabilitation exercises. Their system utilised an RGB-D camera to extract features related to joint trajectories. Expert knowledge was used to select these features. Based on these features, the HSMM provided Clinical Scores for the exercise execution. The HSMM was applicable since the problem adhered to the Markov property; potential future postures are dependent solely on the current posture.

In the realm of wrist rehabilitation, Karime et al. [10] employed a Fuzzy Inference System (FIS) to modify a rehabilitation routine. Their system leveraged sensors to assess patient performance, including accelerometers, gyroscopes, and tracking cameras. The FIS inputs, namely the reach angle, angular velocity, and jerkiness, were derived from the data collected by the sensors. A key advantage of incorporating fuzzy logic in the medical sector is its capacity to manage uncertainty.

2.2 Explainable AI

Explainable Artificial Intelligence (XAI) represents a specific sub-field of Artificial Intelligence (AI), with a primary focus on generating outcomes that humans can comprehend readily [11]. Distinct from the black-box models, XAI models, often referred to as glass-box models, offer a degree of transparency and interpretability that is not typically seen in traditional AI models. Through the provision of explanations for its suggested solutions, XAI fosters a sense of trust in its outputs, thereby paving the way for the broader application of AI across various domains, notably including medicine [12]. It is crucial, however, to acknowledge the existing trade-off between performance and explainability in the AI method employed when designing an XAI model [13]. This balance underscores the challenges inherent in developing models that are both effective and easily interpretable.

Moreover, the categorisation of XAI methodologies can be accomplished based on the approach to interpretation. The interpretation could be intrinsic, meaning it is integral to the AI model, or it could be posthoc, denoting an additional step undertaken on the model. Similarly, interpretation can be global, applying to the model's overall logic, or local, pertaining only to a specific decision for a given instance. Lastly, interpretations may either be model-specific or agnostic, independent of the particular model employed [14].

Gandolfi et al. [15] utilised Machine Learning (ML), specifically Random Forests (RF), to forecast the functional recovery of patients' upper limbs. Along with developing the RF model, they integrated four distinct XAI methodologies to identify the model's most pertinent features. Given that RFs are not inherently interpretable, the employed XAI techniques are considered post hoc, meaning they are applied after the predictive model has been created to help interpret its predictions.

Prentzas et al. [16] constructed a framework that amalgamates Machine Learning (ML) with symbolic reasoning, applying it specifically for stroke predictions. Their framework is capable of being implemented on any ML technique that is compatible with rule-generation algorithms, including Random Forests (RFs) and Decision Trees. In a different context, specifically diabetes diagnosis, Settoui et al. [17] employed neuro fuzzy c-means classifiers. These classifiers represent a fusion of a fuzzy c-means classifier, which offers high interpretability but lacks the capacity for training, and neural networks, which, while trainable, lack inherent interpretability. This combination facilitates the automatic adjustment of fuzzy rules by representing them in a neural network.

3 Architecture

3.1 Foundations

Figure 1 offers a comprehensive architectural overview of the solution proposed in this research work. As can be seen, two applications are interconnected through the cloud infrastructure. The therapist's application facilitates patient monitoring and incorporates features for user management and the definition of rehabilitation routines. In contrast, the patient's application enables the patient to independently carry out rehabilitation exercises at home, adhering to the routine predetermined by the therapist.

Next, key components of the architecture are discussed, considering design aspects that impact the scalability of the proposal when making it accessible for patients requiring remote physical rehabilitation. Section 3.2 introduces the supporting functionality that enables the tracking of the patient's skeleton. Section 3.3 addresses the design of the data architecture which facilitates both the monitoring and evaluation of patient performance, as well as the ability for the system to automatically suggest modifications to the rehabilitation routine previously assigned to the patient. Finally, Sect. 3.4 examines the system's ability to implement such modifications and justify why, through the use of AI.

3.2 Scalable Body Tracking

Remote rehabilitation systems must possess the ability to detect how patients perform rehabilitation exercises in order to evaluate and even automatically

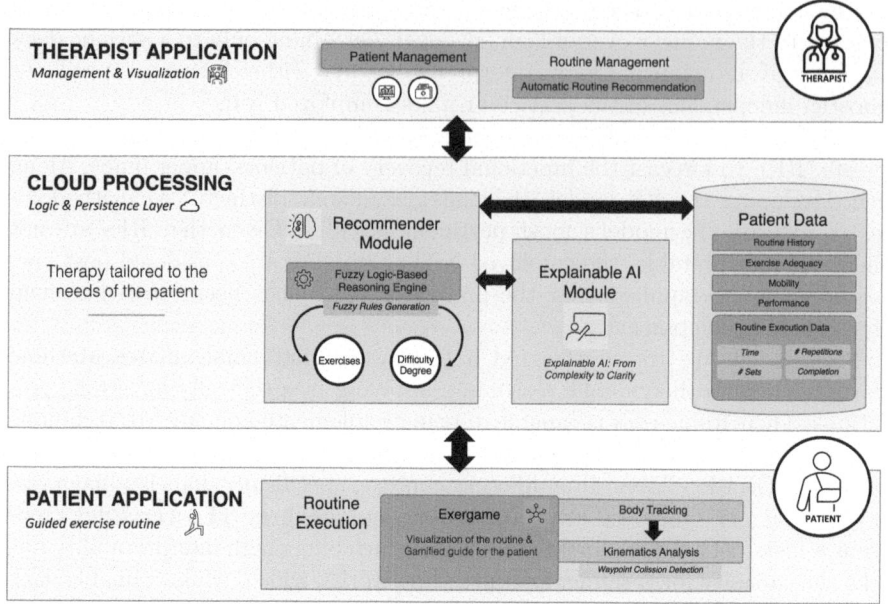

Fig. 1. High-level overview of the proposed architecture.

classify them during a subsequent phase [18]. There are various conventional alternatives for performing body tracking: i) the use of markers or coloured bands placed on the patient themselves, so that body parts can be identified through computer vision; ii) the use of inertial systems comprising sensors placed on the patient's joints; and iii) the use of depth-sensing cameras capable of capturing the 3D spatial position and orientation of the patient's skeleton [19].

Recently, thanks to the proliferation of AI in recent years, systems have emerged that can infer the position of the patient's skeleton by analysing 2D images [20]. This represents a significant advance, as combining a laptop with a standard webcam, or even a mobile phone, makes it possible to have an accessible, low-cost system that effectively resolves the issue of body tracking. One specific example of this kind of tool is MediaPipe[1], which has been utilised in the current proposal.

Figure 2 provides a visual comparison between MediaPipe and XSense, a reference system based on the use of inertial sensors[2]. The latter system employs specific hardware for skeleton tracking and is used in commercial settings, such as in the capture of cinematic animations or as a base for evaluating the performance of elite athletes. On the other hand, MediaPipe offers a significantly more accessible solution, based on the use of machine learning to infer the user's

---

[1] https://developers.google.com/mediapipe.

[2] A comprehensive video comparison is available for the reader at the following URL: https://www.dropbox.com/s/fezecweg9fy3bn6/MediaPipe_VS_XSense.mp4?dl=0.

Fig. 2. Comparison between the results obtained by the MediaPipe library and the XSense inertial capture system when tracking the skeleton.

joint positions from 2D images. In other words, it does not require specific hardware, beyond a standard webcam. As evidenced in Fig. 2, MediaPipe is capable of detecting and appropriately positioning the upper limb joints in 2D space. It can also infer whether a joint is in front or behind (Z coordinate) the user's hip. Opting for a tool like MediaPipe is particularly intriguing when looking to offer accessible, low-cost solutions that can serve as a foundation for the development of remote rehabilitation systems.

3.3 Data Architecture Design

Our proposal relies on a straightforward yet enabling data architecture that allows the storage of how patients perform physical rehabilitation, based on the routine previously assigned by their therapist. Notably, it automatically analyses their performance so that the system can recommend modifications to this routine. These recommendations, artificially generated by the system, may be accepted or rejected by the therapist. At this stage, the explanations that the system is capable of offering become crucial before the therapist makes the final decision. In this regard, our proposal is grounded in the concept of XAI, previously introduced, to facilitate the work of the therapist.

Figure 3 displays the designed data entities, along with their relationships, to facilitate automatic information analysis by the system. One of the fundamental entities is 'routine_execution_data', used to store the detailed information associated with the patient's routine execution. On the other hand, the 'patient_difficulty_data' entity serves to record the patient's progression based on the current difficulty of the assigned routine. This entity is closely related to the 'routine_adjusted_difficulty' entity, which stores information about automatic modifications associated with a specific routine along with explanations as to why the system recommends such modifications.

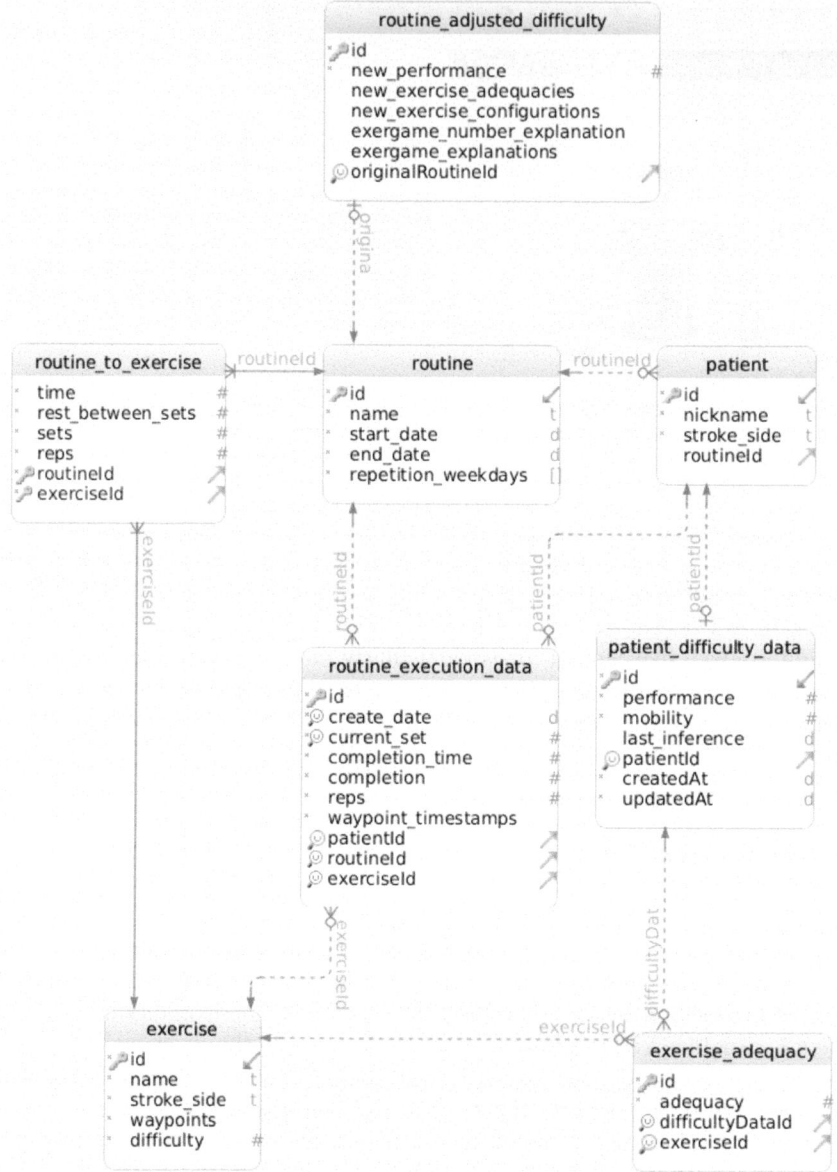

Fig. 3. Entity-relationship diagram that support the data architecture design.

On the other hand, Fig. 4 presents the application used by the therapist to define rehabilitation routines, based on the data architecture previously introduced. As can be observed, a routine comprises a sequenced arrangement of exercises. Each exercise is associated with a specific number of sets, organised into repetitions. Additionally, the therapist has the ability to stipulate the maximum time allocated for the completion of an exercise as well as the rest duration between each set.

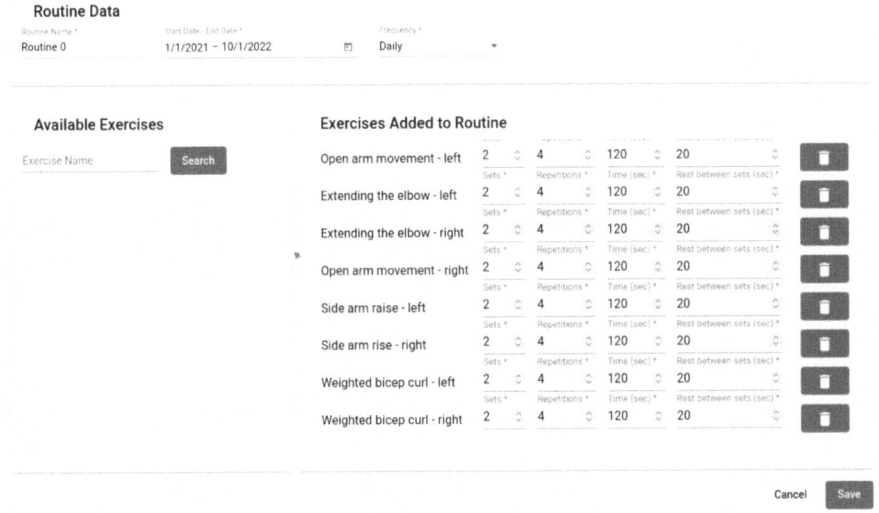

Fig. 4. Therapist editing the patient's routine.

3.4 Automatic Routine Recommendations

Figure 5 presents a sequence diagram that details the interactions between the system actors when the system modifies a rehabilitation routine automatically. As evident from the diagram, the standard workflow involves the patient autonomously carrying out their rehabilitation routine via the patient application, which then uploads the data onto the cloud server.

Moreover, the therapist has the ability to request an automatic modification of the routine, based on the patient's performance and the internal inference module within the therapist's application. However, the therapist always has the final say on accepting the suggested automatic modification of the system. If accepted, this modification is then reflected on the server so that the patient's application can also update it for user interaction.

Deploying the server-side of our proposal in the cloud has several significant advantages. These benefits, which are listed next, are related with the idea of democratising the access to rehabilitation systems based on technology: i) Scalability: cloud platforms are inherently scalable, allowing for the accommodation of varying demand levels. For a remote rehabilitation system, this means that as the number of patients increases or decreases, the system can dynamically

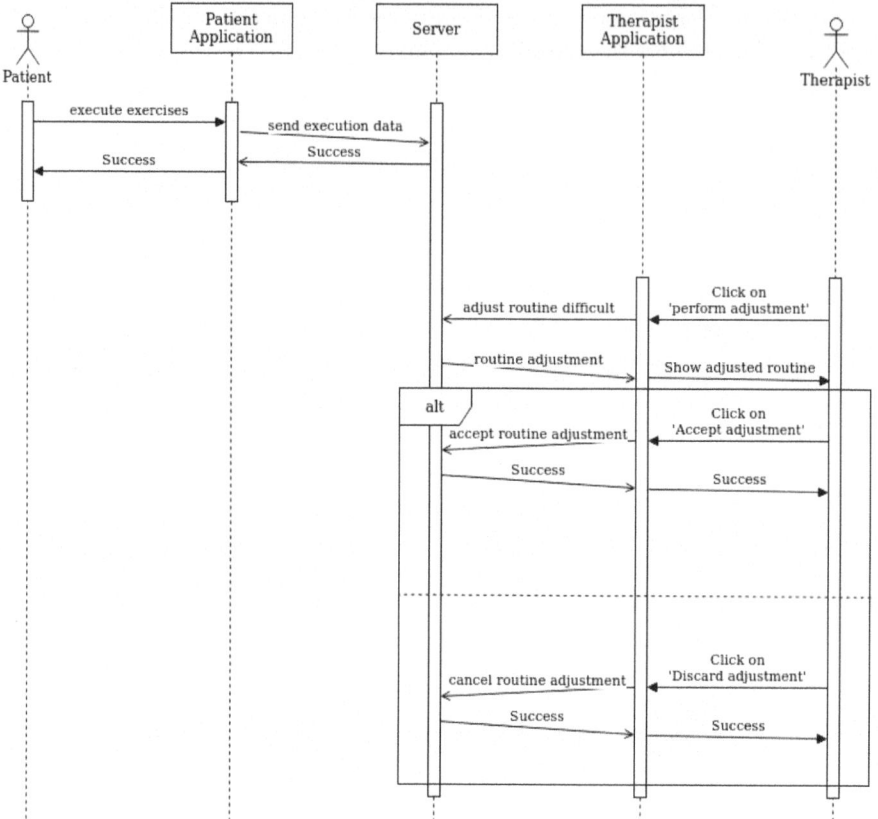

Fig. 5. High-level sequence diagram of the adjustment process.

adjust resources to meet demand, thereby optimising operational costs and performance. ii) Accessibility: with the server-side deployed in the cloud, the system can be accessed anytime and from anywhere, given a reliable internet connection. This ensures uninterrupted provision of services to both therapists and patients. iii) Data Security and Compliance: cloud service providers offer robust security measures to protect stored data, which is essential when dealing with sensitive health information. They also comply with necessary regulations and standards, such as the General Data Protection Regulation (GDPR) in Europe. iv) Collaboration: a cloud-based system enables easy collaboration between therapists, and between therapists and patients. It can provide a unified view of a patient's progress, regardless of the number of therapists working with that patient. v) Reliability and Uptime: cloud-based systems usually have high availability and uptime, ensuring that the rehabilitation service is always available to the patients and therapists. The systems also have redundancy measures in place to prevent data loss. vi) Cost-Effectiveness: deploying server-side systems on the cloud typically follows a pay-as-you-go model, which can be more cost-effective than maintaining on-premise servers. You only pay for the resources you use and you save on infrastructure costs and maintenance. vii) Automatic Updates: the server-side software can be updated or patched automatically, ensuring that the system always uses the most recent and secure version.

Finally, Fig. 6 illustrates the information associated with an example of a system-generated recommendation. The proposed new routine can be seen at the top. At the bottom, the reasons why the system suggests this modification are displayed. For instance, the patient's increased mobility justifies a slight increase in the number of repetitions per set. Likewise, the previous configuration regarding the number of sets and repetitions implies a slight increase in the number of sets.

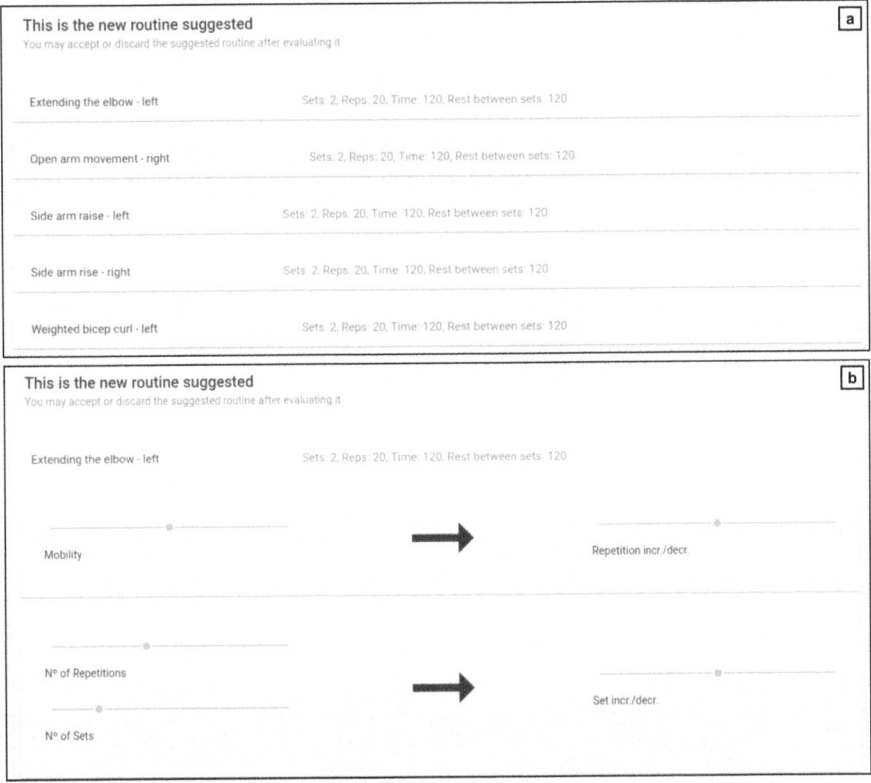

Fig. 6. Automatic recommendation generated by the system. a) The new routine, automatically modified by the system depending on the previously assigned routine and the patient's performance. b) Visual representation of the information used by the system to justify the recommendation.

At this point, it is worth emphasising that the functionality of the automatic recommendations module relies upon the use of Fuzzy Logic [21]. This mathematical approach contributes significantly to enhancing the scalability and interpretability of the system. A noteworthy benefit of integrating fuzzy logic within a medical context is its competency in managing ambiguity and uncertainty. Fuzzy Logic, a specific form of infinite-valued logic, can handle truth values that span from 0 to 1 [22].

Created by Lofti Zadeh in 1965, the primary objective of Fuzzy Logic is to facilitate computation with words [23]. A significant benefit of fuzzy logic is its inherent tolerance for uncertainty, a pivotal capability within our proposal. Considering the imprecise nature of joint tracking and the improbability of dealing with precise data about the patient at all times, the system's ability to address uncertainty is integral to its overall functionality.

The full list of rules and the inference engine used by the automatic recommendations module, coded in R, is available for the reader[3].

4 Patient Application for Remote Rehabilitation

Figure 7 illustrates various screenshots of different views provided by the patient application integrated into the remote rehabilitation system. To facilitate user interaction with the system, it incorporates safety guidelines and recommendations. For instance, Fig. 7.b displays an exercise example, guiding the user on how to move specific body parts for the system to recognise the movements automatically. The system also allows for the execution of specified exercises, as depicted by the exercise selector in Fig. 7.c.

In addition, the system supports functionality related to calibration, considering variations in individual dimensions, as shown in Fig. 7.d. This feature gathers information about the user's skeletal structure and ensures it is properly positioned prior to initiating rehabilitation exercises.

Calibration is essential for a rehabilitation system based on computer vision because it ensures that the system is accurately tracking the patient's movements. This is important for a number of reasons, including accuracy, safety and comfort. Regarding accuracy, if the system is not calibrated correctly, it may not be able to track the patient's movements accurately. This could lead to inaccurate measurements of the patient's progress, which could in turn lead to a less effective rehabilitation program. Regarding safety, if the system is not calibrated correctly, it may not be able to detect potential hazards in the patient's environment. This could put the patient at risk of injury. Finally, regarding comfort, if the system is not calibrated correctly, it may be uncomfortable for the patient to use. This could make it less likely that the patient will continue to use the system, which could impact their progress.

An exercise example is shown in Fig. 7.e, wherein the user is required to laterally move their right arm whilst keeping the elbow fixed. The spheres include numbers that indicate the order in which the user should pass a specific joint over them. In this instance, the joint is the right wrist, represented by the system

---

[3] https://www.esi.uclm.es/www/dvallejo/CAIP/DSS_inference_system.R.

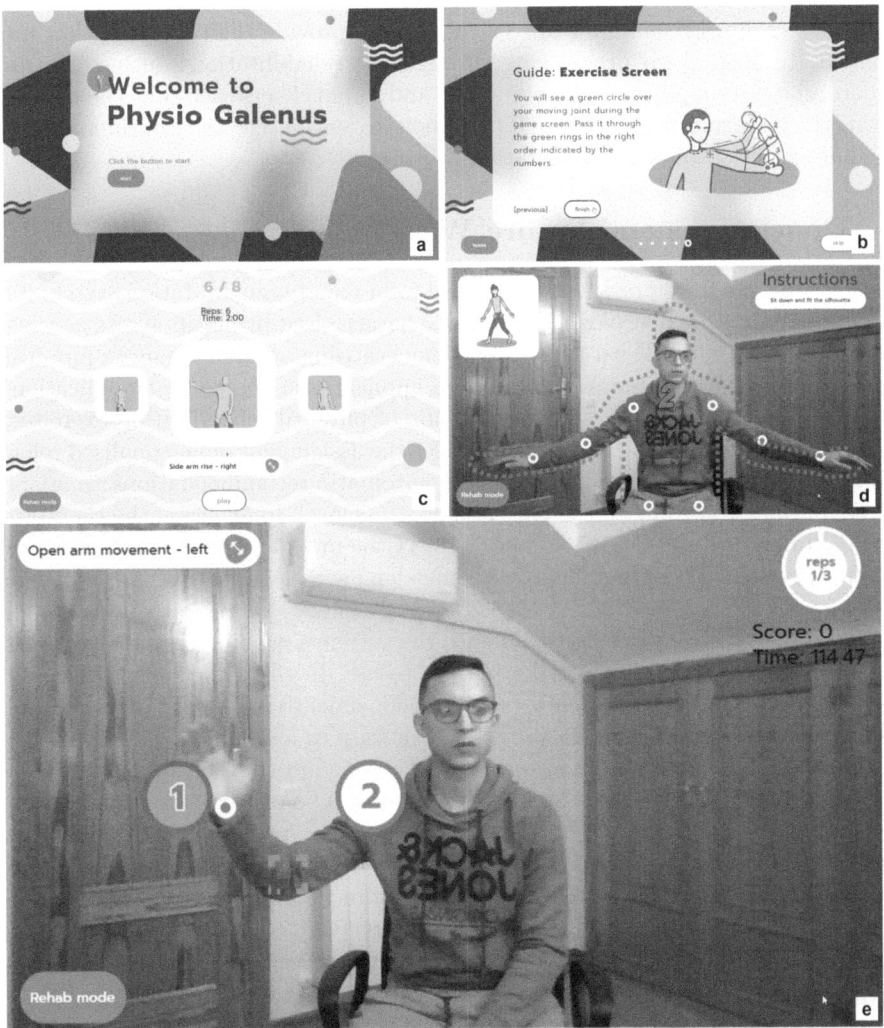

Fig. 7. Screenshots taken from the patient's application. a) Presentation screen. b) Tip sample on how to use the application. c) Selection mode. d) Calibration screen. e) Main screen for performing physical rehabilitation.

with a small white circle with a green centre. The system can also identify the joint to be fixed, in this case, the right elbow, marked with a green cross. This cross should be placed over the pink square, guiding the user to keep his right elbow fixed over the pink marker at all times. This constraint enables the system to guide the user in correctly executing a rehabilitation exercise.

As can be seen, the system displays information about the exercise the user is currently performing (top left of Fig. 7.e,) and information associated with the progress of the rehabilitation work in relation to the assigned routine (top right of Fig. 7.e,). It is important to continually provide the patient with a use context, to keep them progressing through the rehabilitation routine. In this regard, the system plays small animations and sound effects each time the patient correctly performs a repetition. The aim is to encourage the continuous and regular execution of the rehabilitation routine.

5 Conclusions and Future Work

In this work, we have presented a proposal for a remote rehabilitation system for patients affected by neurological diseases, characterised primarily by its accessibility and scalability. With a focus on democratising access to remote physical rehabilitation, this system comprises two applications: the patient application, which allows the recognition and evaluation of physical rehabilitation exercises, and the therapist application, which enables the assignment of personalised rehabilitation routines and the generation of automatic recommendations to adapt these routines according to the patient's progress level. In all cases, the therapist has the final responsibility to accept or reject the modifications proposed by the artificial system. These modifications come along with justifications, infusing the system with elements relating to XAI.

Currently, we are working to enhance the system's usability, ensuring it is as straightforward as possible for the patient to use. In this regard, it is essential to customise the system for each patient independently, as their condition and the exercises they can perform vary from one case to another.

On the other hand, we are also exploring how to incorporate this explainability capability offered by the system for the patient's benefit. For example, we are designing additional functionality so that the system can detect compensations made by the patient when they are unable to perform an exercise. In addition to helping correct these undesired positions, we aim for the system to be capable of explaining how and why they should be corrected.

References

1. Silberberg, D., Anand, N.P., Michels, K., Kalaria, R.N.: Brain and other nervous system disorders across the lifespan–global challenges and opportunities. Nature **527**(7578), 151–154 (2015)
2. Feigin, V.L., Norrving, B., Mensah, G.A.: Global burden of stroke. Circ. Res. **120**(3), 439–448 (2017)
3. Dee, M., Lennon, O., O'Sullivan, C.: A systematic review of physical rehabilitation interventions for stroke in low and lower-middle income countries. Disabil. Rehabil. **42**(4), 473–501 (2020)
4. González-González, C.S., Toledo-Delgado, P.A., Muñoz-Cruz, V., Torres-Carrion, P.V.: Serious games for rehabilitation: gestural interaction in personalized gamified exercises through a recommender system. J. biomed. inf. (Elsevier) **97**, 103266–103285 (2019)
5. Anton, D., Berges, I., Bermúdez, J., Goñi, A., Illarramendi, A.: A telerehabilitation system for the selection, evaluation and remote management of therapies. Sensors (MDPI) **18**(5), 1459–1469 (2018)

6. Karime, A., Mohamad, E., Alja'Am, J.M., El Saddik, A., Gueaieb, W.: A fuzzy-based adaptive rehabilitation framework for home-based wrist training. IEEE Trans. Instrum. Meas. (IEEE) **63**(1), 135–144 (2013)
7. Han, J., Yang, S., Xia, L., Chen, Y.: Deterministic adaptive robust control with a novel optimal gain design approach for a fuzzy 2DOF lower limb exoskeleton robot system. IEEE Trans. Fuzzy Syst. (IEEE) (2021)
8. Pinto, J.F.; Carvalho, H.R. et al. Adaptive gameplay and difficulty adjustment in a gamified upper-limb rehabilitation. In: 2018 IEEE 6th International Conference on Serious Games and Applications for Health(2018)
9. Capecci, M., Ceravolo, M.G., et al.: A Hidden Semi-Markov Model based approach for rehabilitation exercise assessment. J. Biomed. Inform. **78**, 1–11 (2018)
10. Karime, A., Eid, M., Alja, J.M., Saddik, A., Gueaieb, W.: A fuzzy-based adaptive rehabilitation framework for home-based wrist training. IEEE Trans. Instrum. Meas. **63**(1), 135–144 (2014)
11. Ahmed, I., Jeon, G., Piccialli, F.: From artificial intelligence to explainable artificial intelligence in industry 4.0: a survey on what, how, and where. IEEE Trans. Ind. Inf. **18**(8), 5031–5042 (2022)
12. Loh, H.W., Ooi, C.P., et al.: Application of explainable artificial intelligence for healthcare: a systematic review of the last decade (2011–2022). Comput. Methods Programs Biomed. **226**, 107161 (2022)
13. Gunning, D., Stefik, M., Choi, J., et al.: XAI: explainable artificial intelligence. Sci. Robot. **4**(37), 7120 (2019)
14. Payrovnaziri, S.N., Chen, Z., Rengifo-Moreno, P., et al.: Explainable artificial intelligence models using real-world electronic health record data: a systematic scoping review. J. Am. Med. Inform. Assoc. **27**(7), 1173–1185 (2020)
15. Gandolfi, M., Ilaria, B.G., Pavan, R.G., et al.: eXplainable AI allows predicting upper limb rehabilitation outcomes in sub-acute stroke patients. IEEE J. Biomed. Health Inform. **27**(1), 263–273 (2022)
16. Prentzas, N., Nicolaides, A., Kyriacou, E., Kakas, A., Pattichis, C.: Integrating machine learning with symbolic reasoning to build an explainable AI model for stroke prediction. In: 2019 IEEE 19th International Conference on Bioinformatics and Bioengineering (BIBE), pp. 817–821 (2019)
17. Settouti, N., Chikh, M.A., Saidi, M.: Generating fuzzy rules for constructing interpretable classifier of diabetes disease. Australas. Phys. Eng. Sci. Med. **35**(3), 257–270 (2012)
18. Milosevic, B., Leardini, A., Farella, E.: Kinect and wearable inertial sensors for motor rehabilitation programs at home: State of the art and an experimental comparison. Biomed. Eng. Online **19**, 1–26 (2020)
19. Wang, L., Liu, J., Lan, J.: Feature evaluation of upper limb exercise rehabilitation interactive system based on kinect. IEEE Access **7**, 165985–165996 (2019)
20. Hellsten, T., Karlsson, J., Shamsuzzaman, M., Pulkkis, G.: The potential of computer vision-based marker-less human motion analysis for rehabilitation. Rehabil. Process Outcome **10**, 11795727211022330 (2021)
21. Zadeh, L.A.: Fuzzy logic = computing with words. IEEE Trans. Fuzzy Syst. **4**(2), 103–111 (1996)
22. Pelletier, F.J.: Metamathematics of fuzzy logic. Bull. Symbolic Logic **6**(3), 342–346 (2000)
23. Mendel, J., Zadeh, L., et al.: What computing with words means to me. IEEE Comput. Intell. Mag. **5**(1), 20–26 (2010)

Democratization of AI and Governance

Democratization is a Process, not a Destination: Operationalizing Ethics and Democratization in a Cyberinfrastructure for AI Project

Sadia Khan[1]( ), Alfonso Morales[2], and Beth Plale[1]

[1] Indiana University, Bloomington, IN 47405, USA
khanso@iu.edu, plale@indiana.edu
[2] University of Wisconsin, Madison, WI 53706, USA
morales1@wisc.edu

Abstract. In the space of ethical AI, ambiguities abound. Lofty goals such as ethical alignment and systematized fairness are difficult to parameterize, and therefore difficult to achieve. Democratization of AI adds more alignment and operational challenges, but those challenges have not been well articulated or documented as an operational matter. In this paper we present a case study of a large-scale AI research project and discuss processes taken to operationalize ethical and democratizing AI, and the frictions that are inherent in this context. In particular, we discuss challenges associated with a distributed workforce and a mindset focused on research and development of cyberinfrastructure (CI) for AI, both common and seemingly innocuous, but frictious to AI democratization. We find that governance is an important link that holds ethical AI and AI democratization together and moreover that the procedural aspect of governance means that democratization, like democracy, is a process. In telling the measures undertaken in the ICICLE project towards democratization of AI, we contribute a better understanding of the practical challenges of democratizing AI that are particular to infrastructure projects and suggest possible ways for advancing democratization efforts.

Keywords: AI democratization · Cyberinfrastructure · Operationalization

1 Introduction

In the space of ethical Artificial Intelligence (AI), ambiguities abound. Lofty goals such as ethical alignment and systematized fairness are difficult to parameterize, and therefore difficult to achieve. Definitionally, democratization might seem less ambiguous than normative principles like fairness, owing to its roots in political and socio-economic development, but when applied as a goal of artificial intelligence, the clarity of the term seems to evaporate. In this paper we present a case study of a large, federally funded convergence research project in cyberinfrastructure (CI) and describe the challenges of operationalizing a commitment to "democratizing AI." Transplanted from another discipline and vernacular, democratization has not been well theorized in AI literature, much

© ICST Institute for Computer Sciences, Social Informatics and Telecommunications Engineering 2024
Published by Springer Nature Switzerland AG 2024. All Rights Reserved
M. Ziosi et al. (Eds.): CAIP 2023, LNICST 591, pp. 29–45, 2024.
https://doi.org/10.1007/978-3-031-71304-0_3

less studied in situ. As a research project oriented towards cyberinfrastructure, the institute we describe in this case study presents unique insights and learning opportunities. Indeed, because infrastructure is support for other activities, and rarely a goal in and of itself [1, 2], is invisible until breakdown [2], and shapes habits, communities, and environments beyond a single event [2, 3], it is uniquely challenging for democratization, a process itself contingent on individuals' preferences, histories, and beliefs.

In the following sections we discuss these operational issues at the intersection of AI ethics, democratization, and cyberinfrastructure and how they have manifested in the ICICLE research institute (Institute) given the particularities of convergence research. Specifically, we scope the problem of democratizing AI through two main challenge areas: challenges associated with the context of a convergence research institute working to develop intelligent cyberinfrastructure for AI (discussed in section two) and the legibility of AI democratization as a normative issue (discussed in section three). Section four discusses the efforts taken to overcome these challenges to AI democratization from the first two years of work on democratization efforts in the Institute and section five concludes with insights about the operationalization and systematization of AI democratization. In telling the measures undertaken in the Institute towards the democratization of AI, this case study contributes better understanding of the practical challenges of democratizing AI and suggests possible ways of achieving it for others involved in large-scale AI projects.

2 AI Democratization in the Context of Convergence Research: Scoping the Challenge

Global investment in the development of artificial intelligence technologies has burgeoned in recent years, with United States (US) federal government spending on AI contracts alone hitting $3.3 billion USD in fiscal year 2022 [4]. The National Science Foundation (NSF), which funds foundational research, has funded 25 AI Institutes since 2021, each an over $20 million activity focused on foundational and use inspired research in AI, and on educating the next generation of citizens. Intelligent Cyberinfrastructure with Computational Learning in the Environment (ICICLE) [5] is one of those institutes.

Federal agencies will fund large, multimillion dollar projects such as the ICICLE AI Institute based on the recognition that tackling complex problems requires diverse expertise coming together for an extended period to bring a new lens to problems. In such projects, experts frequently form themselves into virtual teams (VTs) which are distributed across research institutions. VTs of geographically and organizationally dispersed co-workers "us[e] a combination of telecommunications and information technologies to accomplish an organizational task" [6]. The ICICLE institute, for example, involves 14 institutions across the US. Each institution has a core leader or Project Investigator, and another 10–15 operational leads who oversee the work of the 120 or so project participants including students, postdoctoral researchers, faculty, and staff. Such is the distributed nature of the AI workforce in institutes like ICICLE.

The ICICLE institute threads together diverse expertise in many domains, including computer science, artificial intelligence, animal ecology, digital agriculture, food science, political and other social sciences, and, like other institutes, incorporates specialized expertise in the practices of broadening participation in computing and workforce development as part of the NSF mandate. Research undertaken by a large team of researchers having diverse backgrounds is known as convergence research, which is "a comprehensive synthetic framework for tackling scientific and societal challenges that exist at the interfaces of multiple fields" [7]. In the case of ICICLE, the Institute aims to carry out research into 1) new software tools and systems in support of AI ("cyberinfrastructure for artificial intelligence" or CI for AI), 2) new AI tools (AI) themselves, and 3) new intelligence within the cyberinfrastructure (AI for CI). Ultimately, ICICLE aims to democratize AI through accessible AI products that are inspired by, but not limited to, problems and opportunities in three use case domains: digital agriculture, animal ecology, and food distribution or smart foodsheds, and are co-developed with expertise in these discipline areas [8]. In short, ICICLE develops components of infrastructure that together will house models to be used for solving specialized problems in agriculture, animal ecology, food systems, and other unknown uses, as well as software for the three use case domains.

A characteristic that is specific to ICICLE relates to generalizability. Research in the Institute will result in software that solves a specific discipline problem, or it will result in software that is part of the infrastructure so invisible to a discipline user. We call the distinction specialization (where the software addresses a narrow problem or need) versus generalization (where the software is part of the infrastructure). This feature is common in the development of infrastructure, but unique to ICICLE among national AI institutes.

In addition to virtual teams and issues of specialization/generalization, a third feature of the Institute is an expectation that it will both generate high quality research that is disseminated through publications within one's core discipline and generate AI and CI tools that function properly. In the case of basic research, software will be written to either embody a research idea or validate it. For its use in discipline settings, the software must undergo a translational process to ready it for use. This could be complex, or as simple as adding processes for handling error. The institute manages a tradeoff between software produced for basic research and software that is useable by discipline scientists.

These three features appear innocuous and obvious. However, we will discuss how their instantiation creates three tensions for AI democratization and AI ethics. First, the distributed workforce as a virtual team poses challenges to accountability and introduces potential hiding places for implicit bias. Second, CI is comprised of software that is generalized for broad use. To achieve AI democratization by measures of social benefit and broader access, multiple communities, timescales, and geographies must be considered. Third, software (AI) research products designed for purposes of validating a scientific theory and only later enter a translation pipeline for actual use, could delay the consideration of ethical and democratizing principles. Post hoc, user-inspired changes may be possible, but they are difficult to retrofit into an AI tool, and more difficult once tools are adopted and habituated.

The first challenge applies generally to all large-scale AI Institutes, while the second and third challenges are particular to the ICICLE AI Institute because of its primary objective of contributing to research infrastructure. So, to practically operationalize AI democratization and follow best practices in ethical AI in ICICLE, essential first steps taken by the Institute's ethics working group and workforce development thrust were to map the problem (as described in this section) and consult prior research to orient democratization in infrastructure of AI and untangle AI democratization and AI ethics.

3 Untangling AI Democratization: Making the Ambiguous Legible

ICICLE has a well-intentioned goal of developing accessible and useable CI for AI, foundational AI, and AI for CI that will benefit society. In the course of writing the proposal to the NSF, project leaders also decided that ICICLE will take a position of "democratizing AI." Together, these two stances about AI access and benefit and AI democratization hold several normative presuppositions: that AI needs normative guidance, that democratizing is desirable, and that greater access to AI tools (alone) meets the normative objectives. To meet normative objectives and measure achievement for reporting purposes, parsing the requirements of AI democratization was necessary. To that end, as part of the process of democratizing AI in the ICICLE project, we went about probing these three assumptions and their relationship to a CI research project through an analysis of related work. We orient cyberinfrastructure with the discourse on AI democratization and discuss related work in the following subsections.

3.1 Democratization as Process Versus Destination

Rajendra-Nicolucci [9] defines *infrastructure* simply as the technologies and systems necessary for society to function. Since it supports human activities in such fundamental ways, social order relies on it [2]. It shapes habits, communities, and environments beyond a single event [3], all the while working in the background—invisible until breakdown [10]. About creating infrastructure, Zuckerman says, "infrastructures are things we build so we can build other things," and building infrastructure is not a goal in itself [1]. However, building working, enduring infrastructure *is* a goal.

Cyberinfrastructure describes large-scale information environments that support research practices through computing services and resources, best practices and standards, visualization environments, and people [11, 12]. As infrastructure, it is inherently challenged by "invisible work, complex problems, and the challenges of alignment in the face of breakdowns" [12], and further, is reconstructed by changing technologies and social needs.

Democratization, a political process of transitioning to a more democratic system [13, 14], requires a reorchestration of institutions, including beliefs and expectations about one's rights to good governance, a desire for democratic governance, and trust in those democratic organizations [15, 16]. Any such large-scale changes are iterative and recursive and proceed in fits and starts [17, 18]. In fact, experts and diplomats experienced with policy and governance commonly conclude about democracy-building, that "Democracy is a process" not a destination [19–21]. Because power is distributed and in the hands of people with various interests, democracy is never finished and is subject to breakdown requiring assessment and incremental reconstruction. We will ultimately conclude the same about democratization of AI; it is a process.

Despite the congruity of the two concepts of democratization and (cyber) infrastructure, where both are processes to support human flourishing and have distant and unknown ends, discussion of cyberinfrastructure and the challenges of democratizing CI for AI are understudied and the process of democratization in this context has not been documented, to our knowledge. Documenting the inherent and idiosyncratic challenges is important because doing so makes legible the ways that infrastructure engages "bodies, habits, categories, communities, labor, environments, politics, and identities" [22], which is essential for surfacing the ethical implications of AI and of the democratization of AI.

3.2 Normative Guidance on AI

Over the past decade, awareness of harms stemming from unfair algorithms and biased data has grown through documented instances of inequalities in numerous books [23–25] and in increasingly regular academic and media reports documenting AI bias [26, 27], privacy concerns [28, 29], concerns over autonomy [30–32], and unfair outcomes in policing [33, 34], hiring [35, 36], and more. Public awareness has not fully extended to understanding the social and scientific practices and systemic injustices that enable these outcomes; however, there has been growing recognition in the use-case domains of ICICLE, following documented cases of AI harms. For instance, privacy risks and unfair outcomes are well documented in digital agriculture [37, 38], animal ecology and welfare [39, 40], and smart food systems [41, 42]. [43–45] and others warn of the scientific research industry's pro-innovation bias that assumes that technology developed to improve farm decisions, for instance, flows linearly and unproblematically from designer to farmer and often fails to recognize differences in the expertise of research scholars and technologists from farmers' local, situated, and informal knowledge gained from everyday interactions with the environment. A dominant belief in agricultural studies is that situated knowledge held by farmers should be considered in scientific research projects [43, 46]. [47] stresses the importance of assessing (not assuming) the benefits of AI in smart farming.

As a result of the heightened concern over harms from AI, AI research and policy communities have emerged which focus specifically on AI ethics. Scoping literature reviews of frameworks for guiding the development of ethical AI show a consensus on a core set of ethical principles that are at stake—namely, transparency; fairness and justice; privacy; beneficence and non-malfeasance; responsibility; and trustworthiness [48, 49]. The AI/computer science research communities have advanced on these issues by developing an arsenal of technical and sociotechnical solutions aligned to normative goals (e.g., algorithmically maximizing fairness [50], training models on local devices rather than distributed networks to strengthen privacy [51], generating explanations for computer decisions to encourage trustworthiness [52], incorporating documentation into the development pipeline for accountability and trustworthiness [53, 54]). Such alignment efforts require parameterizing ambiguous principles for application in potentially diverse and unknown settings, which is not only difficult, but insufficient for achieving ethical AI, which needs governance as well [55, 56] (see Sect. 3.3). What's more, democratization as a goal of AI has received less concerted attention, and therefore its legibility as a normative issue is considerably less developed by comparison to ethical AI principles and techniques.

3.3 Democratization, AI Democratization, and the Ethics of Democratization

Though democracy-building dates to about the 5th century BC Greek civilization, the term democratization was popularized by Samuel P Huntington's 1993 book *The Third Wave: Democratization in the Late Twentieth Century* [13]. The praxis of democratization or democratizing in this domain manifests as a policy supporting democracy-building with the objective of building alliance among ideologically like-minded states.

Democratization of artificial intelligence, another manifestation of an objective to democratize, has been used in the context of AI since the late 1970s [57]. Conceptually, the term has taken many forms. In one early view, [58] supposed that natural language processing (NLP) would make information processing easy to learn, usable, and accessible to anyone, and that democratization of this kind would have "great social effects." Another conceptualization linked democratization to education, training, and knowledge transfer; if different cultures and ideas are represented in computing, resulting systems are more intelligent and more democratic, [59] posited. The term was first used in national policy and in relation to research responsibilities in a 2010 report from the Defense Advanced Research Projects Agency (DARPA). For the report, scientists were asked to develop a framework to help think through ethical, legal, and societal impacts of democratized military technologies—those emergent, with rapid rates of progress and low barriers to adoption [60]. In 2012, DARPA commissioned a report from the National Research Council (NRC) on the same topic. The NRC report elevated a specific concern about who gets access to technologies and what they then do with them [61].

The aspects of diverse representation and public engagement are key in this AI democratization narrative. As Gilman points out, democratic engagement helps avert harmful impacts, adds legitimacy to decisions, and improves accountability [62]. Public engagement is similarly a common topic in AI ethics (e.g. businesses need stakeholder engagement, inclusiveness, and informed discourse in order to produce ethical technologies [63]; public participation is differently conceived depending on the type of organization and its priorities [64]; public participation is hampered by shallow media coverage of AI ethics [65]). However, the discussion of public engagement in AI ethics literature often does not explicitly engage with the goal of AI democratization. By contrast, where public participation is expressed as a feature of democratization, there is an implicit recognition that ethical development (through shared involvement in decision making) cannot be trusted to happen through self-governance alone [62, 66, 67]. Organizational governance should implement measures to facilitate the input of stakeholders through both hard and soft laws [62].

The conceptualization of governance as a meta-responsibility [66] is highly relevant within the scope of operationalizing the democratization of AI in an organizational setting. Indeed, Seger et al.'s [68] taxonomy of AI democratization, which was derived from the commitments to democratization expressed by AI companies' labs, includes the democratization of AI governance as one of the four meanings of AI democratization along with democratization of AI use, democratization of AI development, and democratization of AI profits. This taxonomy is helpful for untangling what it means to democratize AI; we will return to it in Sect. 4.1 [69].

As a term, democratization of AI may seem relatively disconnected from democratization as a policy objective, but as processes they share issues of measurement and ethical implications. As [70] emphasizes, the concepts that underpin conceptions of democratization are EuroAmerican, and these conceptualizations become part of a standardized measure of what it means to democratize, to be democratic, or to reach democratization. Democratization and globalization have failed, or worse yet, supplanted local beliefs and desires in many cases because the process of democratization did not account for different histories and cultures [71–73]. Similarly with democratization of AI, it is not an absolute good [68]. It matters how the goals, recommendations, and guidelines of democratization are specified, no matter which manifestation of democratization [74].

Each of the meanings of democratizing AI discussed above actually represents individual goals with respect to AI democratization, and achieving each component has different requirements. Internalizing these needs, and the concerns that ground them, was part of our process of operationalizing the commitment to democratizing AI for ICICLE, and in the next section we discuss the measures we have taken with respect to the particular tensions of cyberinfrastructure and convergence research.

4 Findings on Operationalizing Democratization in ICICLE

From analysis of the theory and practice/process of democratization, it became clear that while AI ethics and democratization share some features (for example benefit/beneficence), they are not coterminous. This means, to advance democratization in the ICICLE Institute requires dedicated work on democratization, some of which is separate from AI ethics. This was surmised from an exercise to map what it means to be democratizing from center to edge. We also determined that to be robust and enduring, operationalizing AI democratization requires work on democratization on multiple fronts related to its multiple meanings. These map onto the challenges described in section two, where challenges for AI democratization arise with the distributed workforce, the generalized nature of cyberinfrastructure, and the need to produce software that validates a scientific theory at one point in time and then later becomes a product for use (see Fig. 1). In the following subsections, we discuss what we have found to be the key challenges for democratizing AI in the contexts of large-scale research projects and cyberinfrastructure. We also document ways we have addressed those challenges in ICICLE—providing an operational vantage that is underrepresented.

4.1 The Institutionalization of Multiple Meanings of Democratization

The ICICLE institute commenced with a mental model that formulating well-intentioned ethical AI (e.g., using practices from privacy, provenance, and explainability) with democratizing goals (e.g., broad outreach to use communities, Work Force Development, training, ease of access) would iterate towards democratizing AI. In the first two years, as AI and CI experts explored research advances in CI and AI components, simultaneously the privacy, accountability, and ethics work group (PrivEthx) and Work Force Development (WFD) research thrusts worked toward understanding the component parts of operationalizing the goal of democratizing AI. A culmination of that research led to

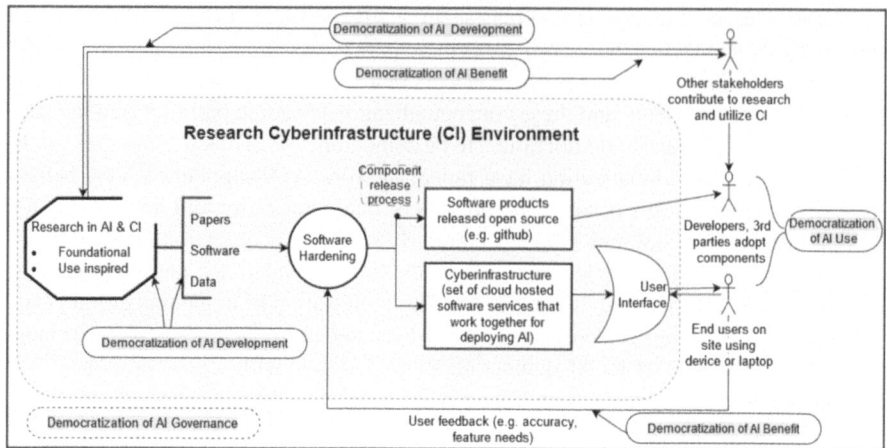

Fig. 1. Democratization Opportunities in the Cyberinfrastructure Environment

the adoption of a conceptualization of AI democratization that includes the democratization of AI use/access, of AI benefit, of AI development, and of AI governance (see Fig. 2). This is similar to [68]'s taxonomy derived from AI companies (see Sect. 3.3), but rearticulates AI profit to be about AI benefit, and defines AI governance more operationally as an adaptation of Seger et al., and according to conceptualizations in other literature.

Democratization of AI Access	Making AI software, models, and information more accessible to a wider range of potential users.
Democratization of AI Benefit	As with fairness, all people should benefit from AI. AI should not harm users or create greater disparity.
Democratization of AI Development	Having a wider range of stakeholders contribute their perspectives and expertise to its design and development process.
Democratization of AI Governance	Facilitating the representation of diverse and conflicting values and perspectives on how people and their actions are governed [69]. Maintaining oversight over the democratization of AI access, benefit, and development.

Fig. 2. Framework for the Democratization of AI

From the beginning, the work proceeded in fits and starts. A consequential example includes a sustained debate about our tagline, "Democratizing AI" which ended in eventually acknowledging that our proposal writing and budgeting process did not fully account for the intense interest that external stakeholders would have in this aspect of our work, which led to reorganizing and additional discussions and research. It also encouraged writing (this paper for instance) about our unique context of democratization.

The process' recursive nature also stems from discipline-diverse virtual teams facing a considerable task in reconciling and harmonizing different vocabularies and understandings, especially on cross-institute topics. So, it took time to surface the sometimes vastly divergent views on what constituted AI ethics and democratization. As a result, a new work group was formed with members from both PrivEthx and WFD. The work group undertook a comprehensive literature review that had the effect both of untangling AI democratization and AI ethics and of serving as a foundation upon which cross-team discussions could begin/start. From this we identified 4 major concepts associated with AI ethics, with democratization of AI as a 5th, separate concept. While others are relevant, for ICICLE as a research cyberinfrastructure project, we focus on the challenges of trustworthiness, privacy, fairness, accountability, and democratization. This essential first step of framing laid important groundwork for operationalizing democratization in ICICLE.

Towards Democratization of AI Development. An early effort was to leverage the ICICLE video series, a series of 3–5 min videos used to raise awareness and understanding. The video series allowed us to reach the ICICLE team with short and clarifying guidance distinguishing AI ethics and democratization. Each video provided accessible context from history and theory about one of the ethics concepts and provided tips for implementing the concepts, such as tips on centering stakeholder needs. The WFD group additionally created training material for a high school summer camp based on the per-concept approach, with each concept accompanied by an interactive exercise. The goal of these efforts was to initiate culture change in scientific practice towards considering ethical implications and incorporating stakeholder perspectives.

Towards Democratization of AI Benefit and Governance. Framing the democratization of AI as being about access, development, benefit, and governance allows us to clarify who is the beneficiary and what is the Institute's obligation to user engagement in the research process. It has become clear over time that user engagement in software being developed to validate a basic research question will take a different form from user engagement in product translation through a translation pipeline. Similarly, user engagement in end-facing AI tools will look different than will engagement in the development of general cyberinfrastructure components. For example, of the three types of beneficiaries depicted in Fig. 1, one benefits from functional and accessible AI and CI, and others benefit from accessible tools that are well suited for their needs. Though, assessment of user needs in the former is complicated by the uncertainty about who are end-users. In both of these cases, if governance were to include guidance from [47] to assess (and not assume) the benefits of AI, the forms that the assessment would take would vary in terms of who assesses, what is assessed, and when in the software lifecycle it is assessed. This has led us to focus our next stage of research on the utility of governing rules for operationalizing democratization.

4.2 Challenges of a Distributed Workforce

We observe that the distributed VT structure poses challenges to AI democratization by introducing potential places in organizational structures that foster inaction and obfuscate implicit bias. For one, we observed that even with having experts in participatory and ethnographic methods on the ICICLE team, those experts say they feel no home in the scientific research thrusts because the virtual research thrusts tend to have homogenous expertise relevant to solving a particular problem, not building understanding. As such, the VT ecosystem has a siloing effect where diverse expertise may get introduced, but not integrated. Relatedly, academic roles are notoriously independent, and scholars often eschew standard operating procedures. Typically, workplace environments in non-VT organizational settings will have standard operating procedures that reconcile individual expectations and experiences and subordinate those to larger organizational or social objectives. The accountability this brings is difficult to achieve in VTs involving multiple universities and competing interests. Indeed, the science of "team" building emerged as a unique discipline to study and establish operating procedures that could reconcile distinct expectations of the disciplines doing convergence research into a coherent whole to advance project objectives across diverse expectations [75].

Towards Democratization of AI Development. Two efforts to address these challenges have focused on organizational restructuring and team member training and team building. First, as an organizational rule, the ongoing restructuring encourages the integration of diverse perspectives on AI development through the inclusion of participatory design researchers into the CI, AI, and privacy research thrusts. Second, similar to the ethics videos described in Sect. 4.1, short allyship videos were designed to serve as a resource and bring awareness to the team about inclusion and implicit bias. A Justice-Equity-Diversity-Inclusion (JEDI) workshop conducted in person with the entire team present provided lessons and exercises for promoting a healthy team. A post-workshop assessment showed increased awareness of the importance of amplifying the voices of the least powerful. Additionally, teambuilding among students and early career professionals is being promoted by a "Next-Gen" affinity group which facilitates opportunities for networking despite physical distance. By these combined efforts, not only should diverse perspectives be integrated, but they should also be heard.

4.3 Challenges of Specialization/Generalization

As discussed in section two, a strong focus of the Institute is on research that results in advances in intelligent cyberinfrastructure. Where some of the software products resulting from cyberinfrastructure research might have benefitted from participatory design, cyberinfrastructure software requirements are often based on factors such as cost, energy usage, etc. As the Institute explores opportunities to push AI into cyberinfrastructure, among its other goals, there is a nonzero risk that various biases and cultural predispositions could exist. Traditional participatory design approaches which have a democratizing effect on benefit are not particularly amenable to this challenge because cyberinfrastructure is not developed to cater to particulars and, as mentioned in 4.2, incorporating participatory design is challenged by virtual teams.

Towards Democratization of AI Benefit. The creation of generalized cyberinfrastructure can be guided by democratization of AI benefit and here, too, the ethics videos can help expand thinking about benefit. However, that challenge cannot be met without larger resocializing efforts (see 4.4 below). On the other hand, a more technical method we leverage to bridge generalization and specialization concerns is model cards [54]. Model cards are machine readable "cards" that contain context about computational models and other software tools, including context about people who create the models, and about processes and data that went into the models. Innovations which add specialized context to general models can help minimize harms and control out of context uses, even when the end user and use is unknown. For example, adding documentation about privacy vulnerabilities and the social and material conditions that a model was tested on, can aid users tremendously in picking a model that will make the most appropriate predictions for their context.

4.4 Challenges of Scientific Research to End Product Progression

The process of scientific discovery and the particular needs of end users are distinct, and not easily reconciled. This is because various incentive structures make the benefits and rewards of discovery altogether different from the benefits to end users. There are competing primary expectations across the Institute. One is that a software tool should foremost contribute to scientific discovery, the other prioritizes deeper engagement with end users or their problems. In the former, the benefit to end-users is frequently considered only vaguely, if at all, in the evaluation of the research product. In this case, benefit is incidentally deprioritized. However, that is the logic of scientific discovery, and such requires sensitizing and resocialization.

Without broader engagement, incomplete representations of human and non-human populations and circumstances are used in core product discovery and development processes. Absent engagement, this is nearly inevitable because how can humans fully represent non-human stakeholders or how can scientists represent all, or at least the most salient features of human populations and livelihoods, given that those are contextual and changing?

Towards Democratization of Benefit and Governance. The ethics videos (described in 4.1) serve a resocializing purpose by working to change habits of mind about responsibility to end users and end products, raise awareness of unintended consequences specific to (invisible) infrastructure, and center stakeholder needs. However, video reminders are clearly not sufficient for operationalizing democratization. We also do several things to bring stakeholders in more materially (i.e., beyond a reminder of their invisible interests). One is the organizational restructuring that gives greater focus on the work of ICICLE scientists who use community-based research methods and other fieldwork. In the restructuring process, all members of ICICLE (leadership, student researchers, staff, etc.) are encouraged to participate and give feedback on reorganizing and redefining priorities. Ideally, the Institute and its products should be more democratic as a result. Too, while not yet operational, we have established a stakeholder roundtable to give users a seat at the table in matters of interest to them.

5 Translating Research into Insights

Recently there has been a growth in research on ethical AI—introducing technical methods and policy frameworks to combat bias in models, data, and institutional structures that perpetuate AI harms. None have been sufficient and all suffer from challenges related to operationalizing ambiguous normative concepts given that, for example, fairness and benefit look different for different users, and, in some cases, users are unknown. This case study has exposed tensions and opportunities and expanded understanding of AI democratization in the understudied context of a research cyberinfrastructure institute. For convergence research, the uncertainty at the nexus of the specialization vs. generalization tension and the research-product vs. translated-end-product progression/tension creates operationalization issues that, while qualitatively unique, have broader implications for AI democratization in the context of infrastructure. Next, we draw some insights and tips from our findings and experiences.

Ethical AI Democratization is not Guaranteed. Democratization is not an absolute good. It carries with it implicit biases about what is good for a society. Because it is not naturally a part of scientific research processes to incorporate such thinking, a concerted effort is required. Specifically, effort should be made to move beyond the notion that AI access is necessarily beneficial and synonymous with democratization. In many cases (e.g. agriculture, medicine, commerce), access to novel technologies have increased disparities and supplanted indigenous knowledge and practices. Incorporating those knowledge systems into AI development and considering the costs and benefits for potential users are vital parts of AI democratization.

Insight 1. To increase the chance that democratization is beneficial, teams that wish to operationalize AI democratization should begin with an expansive conceptualization of democratization that includes human and non-human participants and their needs.

Anticipate that AI Democratization is an Unfolding Process. As we showed in Sect. 4.1, operationalizing a commitment to democratization is a process. At points it led to a breakdown that required rethinking the entire commitment laid out in the original research proposal. By exposing these fits and restarts, others may be encouraged to implement a full commitment to democratization efforts from the beginning; integrate the democratization research team into the other science research areas; and assign resources to the democratization task so that the team is not resource stressed.

Insight 2. The path to AI democratization can be fraught, but making a fully resourced commitment to democratization from the beginning can ease the process and give credibility to the effort.

Building Cyberinfrastructure Demands Some Amount of Seeing the Invisible. Infrastructure is something you build so that others can build other things. With research cyberinfrastructure, the translated end products that result from academic research are difficult to know, and thus, so too are the uses and implications. Building cyberinfrastructure demands some amount of seeing the invisible. For example, in ICICLE, we have acknowledged the concerns laid out by [43–45, 76] about the scientific research industry's pro-science bias which has the potential to dismiss or devalue local knowledge and [77]'s concerns about the potential tradeoff between access/accessibility and privacy. We have made these concerns legible in several ways described in the paper, including advocating for the inclusion of stakeholders.

Insight 3. There must be efforts to build institute culture and awareness around normative issues to encourage seeing beyond the code. It is these norms that will reach the student who is doing purely foundational research work towards generalized cyberinfrastructure.

Normative Guidelines Do Not Replace Governance. Much of what is considered "democratizing" in AI is encompassed in what it means to have ethical AI. This is supported by the literature, and we found it to be true in parsing the operational requirements of AI democratization in ICICLE. One particularity is that, as an operational matter, governance sits as a meta-responsibility in AI democratization, as [62] says, lest democratization be left to self-governance. Moreover, that governance must include diverse perspectives. The breakdowns in ICICLE over operationalizing ICICLE's commitment to democratization demonstrate the frictions that impede self-governance and make more formal rules and procedures necessary for operationalization.

Insight 4. Governance is needed to ensure institutional commitment to the normative goals the institute establishes. Those commitments should encourage accountability to ethical and democratizing AI and come about through democratic processes.

The Translation Pipeline is an Important Conduit to Encourage Culture Change. AI project governance is not typically oriented to ethics. Instead, it is oriented to project objectives, and those are associated with disciplines and their objectives, which do not articulate and defend social goals. When multiple disciplines and organizations are joined in one project, such as ICICLE, the principal problem is the disparate disciplinary and personal perspectives one has with respect to scientific engagement and one's conceptualizations related to commitments to science versus social benefit.

Insight 5. The translation pipeline can be a critical point of intervention; a strategically placed intervention can expand understanding across the entire VT, thus realizing a culture of ethically-grounded interdisciplinary research.

6 Conclusion

This paper has sketched a variety of operational issues at the intersection of AI ethics, democratization, and cyberinfrastructure and how they have manifested in a large-scale, federally funded AI research institute with a tagline of "democratizing AI." Specifically, we describe the challenges for AI democratization that arise from the distributed workforce, the generalized nature of cyberinfrastructure, and the gap between producing software that validates a scientific theory at one point in time and possibly becoming a product for use at a later point in time. Work in the ICICLE institute to address these challenges through workforce development and culture change, governance mechanisms, and stakeholder engagement are ongoing, and the process will continue. As experts, we know from hindsight that a failure to build guidelines and guardrails on AI development at early stages in other domains exacerbated harms when AI was deployed. Bringing the tensions to light, as we have in this paper, makes legible some of the ethical implications of AI democratization in order that others involved in AI projects may better anticipate challenges and iterate towards ethical and democratizing AI technologies.

Acknowledgements. This research is funded in part by US National Science Foundation, Award # 2112606.

References

1. Zuckerman, E.: What Is Digital Public Infrastructure? https://www.journalismliberty.org/publications/what-is-digital-public-infrastructure. (2020)
2. Star, S.L.: The ethnography of infrastructure. Am. Behav. Sci. **43**, 377–391 (1999)
3. Rossiter, N.: Software, Infrastructure, Labor : A Media Theory Of Logistical Nightmares /. Routledge Taylor & Francis Group, New York (2016)
4. Krishan, N.: Federal gov spending on AI hit $3.3B in fiscal 2022: study (2023). https://fedscoop.com/us-spending-on-ai-hit-3-3b-in-fiscal-2022/
5. ICICLE: Intelligent CI with Computational Learning in the Environment,. https://icicle.osu.edu/
6. Townsend, A.M., DeMarie, S.M., Hendrickson, A.R.: Virtual teams: technology and the workplace of the future. Acad. Manag. Perspect. **12**, 17–29 (1998)
7. Council, N.R.: Convergence: Facilitating Transdisciplinary Integration Of Life Sciences, Physical Sciences, Engineering, and Beyond. National Academies Press (2014)
8. ICICLE: Use Inspired Science. https://icicle.osu.edu/about-us/use-inspired-science
9. Rajendra-Nicolucci, C.: Keyword: Accidental Infrastructure. https://publicinfrastructure.org/2023/02/01/keyword-accidental-infrastructure/
10. Star, S.L., Ruhleder, K.: Steps towards an ecology of infrastructure: complex problems in design and access for large-scale collaborative systems. In: Proceedings of the 1994 ACM conference on Computer supported cooperative work, pp. 253–264 (1994)
11. Stewart, C.A., Simms, S., Plale, B., Link, M., Hancock, D.Y., Fox, G.C.: What is cyberinfrastructure. In: Proceedings of the 38th Annual ACM SIGUCCS Fall Conference: Navigation and Discovery, pp. 37–44 (2010)
12. Bowker, G.C., Baker, K., Millerand, F., Ribes, D.: Toward Information Infrastructure Studies: Ways of Knowing in a Networked Environment. In: Hunsinger, J., Klastrup, L., Allen, M. (eds.) International Handbook of Internet Research, pp. 97–117. Springer, Netherlands, Dordrecht (2010)
13. Huntington, S.P.: The Third Wave: Democratization in the Late Twentieth Century. University of Oklahoma Press (1993)
14. Baloyra, E.A.: Comparing new democracies: transition and consolidation in mediterranean Europe and the southern Cone. Routledge (2019)
15. Dahl, R.A.: The problem of civic competence. J. Democracy. **3**, 45 (1992)
16. Kadivar, M.A.: Mass mobilization and the durability of new democracies. Am. Sociol. Rev. **83**, 390–417 (2018). https://doi.org/10.1177/0003122418759546
17. Walker, W.E., Rahman, S.A., Cave, J.: Adaptive policies, policy analysis, and policy-making. Eur. J. Oper. Res. **128**, 282–289 (2001)
18. North, D.C.: Institutions. Cambridge University Press, Institutional Change and Economic Performance (1990)

19. McKinnon, D.: Democracy not a destination but a journey. https://thecommonwealth.org/news/democracy-not-destination-journey-don-mckinnon
20. Przeworski, A.: Transition to Capitalist Democracy as Class Compromise. Classes and Elites in Democracy and Democratization. New York: Garland Publishing. 128–133 (1997)
21. Lappé, F.M.: Democracy's Edge : Choosing to Save Our Country by Bringing Democracy to life /. Jossey-Bass, San Francisco : (c2006.)
22. Rich, K.M., Rizzuto, N.M., Zieger, S.: The Aesthetic Life of Infrastructure: Race, Affect, Environment. Northwestern University Press (2022)
23. Eubanks, V.: Automating inequality : How High-Tech Tools Profile, Police, and Punish the Poor. St. Martin's Press (2018)
24. O'Neil, C.: Weapons of math destruction : how big data increases inequality and threatens democracy /. Crown, New York, NY (2016)
25. D'Ignazio, C., Klein, L.F.: Data feminism. MIT Press, Cambridge, MA (2020)
26. Buolamwini, J., Gebru, T.: Gender shades: intersectional accuracy disparities in commercial gender classification. Proc. Mach. Learn. Res. **81**, 1–15 (2018)
27. Bolukbasi, T., Chang, K.-W., Zou, J.Y., Saligrama, V., Kalai, A.T.: Man is to computer programmer as woman is to homemaker? Debiasing word embeddings. Adv. neural inf. Proc. syst. **29** (2016)
28. Acquisti, A., Gross, R., Stutzman, F.D.: Face recognition and privacy in the age of augmented reality. J. Priv. Confidentiality. **6**, 1 (2014)
29. Fuchs, C.: The political economy of privacy on Facebook. Telev. New Media **13**, 139–159 (2012)
30. Susser, D., Roessler, B., Nissenbaum, H.: Technology, autonomy, and manipulation. Internet Policy Rev. **8**(2) (2019). https://doi.org/10.14763/2019.2.1410
31. Rastogi, A., Nygard, K.: Trust and security in intelligent autonomous systems. Int. J. Comput. Their Appl. **26**, 22–29 (2019)
32. Rubel, A., Castro, C., Pham, A.: Algorithms and Autonomy: The Ethics of Automated Decision Systems. Cambridge University Press (2021)
33. Lum, K., Isaac, W.: To predict and serve? Significance **13**, 14–19 (2016)
34. Dressel, J., Farid, H.: The accuracy, fairness, and limits of predicting recidivism. Sci. adv. **4**(1), eaao5580 (2018)
35. Dastin, J.: Insight - Amazon scraps secret AI recruiting tool that showed bias against women (2018). https://www.reuters.com/article/amazon-com-jobs-automation-idINKCN1MK0AH
36. Yam, J., Skorburg, J.A.: From human resources to human rights: impact assessments for hiring algorithms. Ethics Inf. Technol. **23**, 611–623 (2021). https://doi.org/10.1007/s10676-021-09599-7
37. Sparrow, R., Howard, M., Degeling, C.: Managing the risks of artificial intelligence in agriculture. NJAS Impact Agric. Life Sci. **93**, 172–196 (2021). https://doi.org/10.1080/27685241.2021.2008777
38. Janc, K., Czapiewski, K., Wójcik, M.: In the starting blocks for smart agriculture: the internet as a source of knowledge in transitional agriculture. NJAS Wageningen J. Life Sci. **90**, 100309 (2019). https://doi.org/10.1016/j.njas.2019.100309
39. Hagendorff, T.: The Ethics of AI Ethics: an evaluation of guidelines. Mind. Mach. **30**, 99–120 (2020). https://doi.org/10.1007/s11023-020-09517-8
40. Schewe, R.L., Stuart, D.: Diversity in agricultural technology adoption: how are automatic milking systems used and to what end? Agric. Hum. Values **32**, 199–213 (2015). https://doi.org/10.1007/s10460-014-9542-2
41. Demarsh, N., Morales, A.: The practical ethics of urban agriculture. In: Raja, S., Born, B., Caton-Campbell, M., and Morales, A. (eds.) The Food System Scholarship of Jerome Kaufman. Springer, Toronto Ontario Canada

42. Jackson, R.: Unpacking the ethics of food sustainability: health, harmony and beyond. https://www.nuffieldbioethics.org/blog/unpacking-ethics-food-sustainability-health-harmony
43. Rose, D.C., Morris, C., Lobley, M., Winter, M., Sutherland, W.J., Dicks, L.V.: Exploring the spatialities of technological and user re-scripting: the case of decision support tools in UK agriculture. Geoforum **89**, 11–18 (2018). https://doi.org/10.1016/j.geoforum.2017.12.006
44. Holloway, L., Bear, C., Wilkinson, K.: Robotic milking technologies and renegotiating situated ethical relationships on UK dairy farms. Agric. Hum. Values **31**, 185–199 (2014). https://doi.org/10.1007/s10460-013-9473-3
45. Morris, C.: Environmental knowledges and expertise. In: International Encyclopedia of Geography, pp. 1–8. John Wiley & Sons, Ltd (2017)
46. Oliver, D.M., Fish, R.D., Winter, M., Hodgson, C.J., Heathwaite, A.L., Chadwick, D.R.: Valuing local knowledge as a source of expert data: farmer engagement and the design of decision support systems. Environ Model Softw. **36**, 76–85 (2012). https://doi.org/10.1016/j.envsoft.2011.09.013
47. Shepherd, M., Turner, J.A., Small, B., Wheeler, D.: Priorities for science to overcome hurdles thwarting the full promise of the 'digital agriculture' revolution. J. Sci. Food Agric. **100**, 5083–5092 (2020). https://doi.org/10.1002/jsfa.9346
48. Jobin, A., Ienca, M., Vayena, E.: The global landscape of AI ethics guidelines. Nat. Mach. Intell. **1**, 389–399 (2019). https://doi.org/10.1038/s42256-019-0088-2
49. Fjeld, J., Achten, N., Hilligoss, H., Nagy, A., Srikumar, M.: Principled artificial intelligence: mapping consensus in ethical and rights-based approaches to principles for AI. SSRN J. (2020). https://doi.org/10.2139/ssrn.3518482
50. Martinez, N., Bertran, M., Sapiro, G.: Minimax pareto fairness: a multi objective perspective. In: Proceedings of the 37th International Conference on Machine Learning, pp. 6755–6764 (2020)
51. Li, T., Sahu, A.K., Talwalkar, A., Smith, V.: Federated learning: challenges, methods, and future directions. IEEE signal process. mag. **37**, 50–60 (2020)
52. Ribeiro, M.T., Singh, S., Guestrin, C.: "Why Should I Trust You?": Explaining the Predictions of Any Classifier. arXiv:1602.04938 [cs, stat]. (2016)
53. Barker, M., et al.: FeedbackLogs: Recording and incorporating stakeholder feedback into machine learning pipelines. http://arxiv.org/abs/2307.15475 (2023)
54. Mitchell, M., et al.: Model cards for model reporting. In: Proceedings of the Conference on Fairness, Accountability, and Transparency, pp. 220–229 (2019)
55. Eitel-Porter, R.: Beyond the promise: implementing ethical AI. AI Ethics. **1**, 73–80 (2021). https://doi.org/10.1007/s43681-020-00011-6
56. Mittelstadt, B.: Principles alone cannot guarantee ethical AI. Nat. Mach. Intell. **1**, 501–507 (2019). https://doi.org/10.1038/s42256-019-0114-4
57. Olazaran, M.: A Sociological study of the official history of the perceptrons controversy. Soc. Stud. Sci. **26**, 611–659 (1996). https://doi.org/10.1177/030631296026003005
58. Gurstein, M.: Social impacts of selected artificial intelligence applications: the Canadian context. Futures **17**, 652–671 (1985)
59. Gill, K.S.: Artificial Intelligence and Social Action: Education and Training. In: Göranzon, B., Josefson, I. (eds.) Knowledge, Skill and Artificial Intelligence, pp. 77–91. Springer, London (1988)
60. Chameau, J.L., Ballhaus, W.F., Lin, H.S.: Emerging and readily available technologies and national security : a framework for addressing ethical, legal, and societal issues. National Academies Press, Washington, DC (2014)
61. Fischhoff, B.: Ethical and social issues in military research and development. Telos **2014**, 150–154 (2014). https://doi.org/10.3817/1214169150
62. Gilman, M.: Democratizing AI: Principles for Meaningful Public Participation. Data Soc. (2023)

63. Buhmann, A., Fieseler, C.: Towards a deliberative framework for responsible innovation in artificial intelligence. Technol. Soc. **64**, 101475 (2021). https://doi.org/10.1016/j.techsoc.2020.101475
64. Schiff, D., Borenstein, J., Biddle, J., Laas, K.: AI Ethics in the Public, Private, and NGO Sectors: A Review of a Global Document Collection (2021). https://www.techrxiv.org/articles/preprint/AI_Ethics_in_the_Public_Private_and_NGO_Sectors_A_Review_of_a_Global_Document_Collection/14109482/1
65. Ouchchy, L., Coin, A., Dubljević, V.: AI in the headlines: the portrayal of the ethical issues of artificial intelligence in the media. AI Soc. **35**, 927–936 (2020). https://doi.org/10.1007/s00146-020-00965-5
66. Buhmann, A., Fieseler, C.: Deep learning meets deep democracy: deliberative governance and responsible innovation in artificial intelligence. Bus. Ethics Q. **33**, 146–179 (2023). https://doi.org/10.1017/beq.2021.42
67. Voegtlin, C., Scherer, A.G.: Responsible innovation and the innovation of responsibility: governing sustainable development in a globalized world. J. Bus. Ethics **143**, 227–243 (2017). https://doi.org/10.1007/s10551-015-2769-z
68. Seger, E., Ovadya, A., Garfinkel, B., Siddarth, D., Dafoe, A.: Democratising AI: Multiple Meanings, Goals, and Methods. http://arxiv.org/abs/2303.12642 (2023)
69. Seger, E.: What Do We Mean When We Talk About "AI Democratisation"? | GovAI Blog (2023). https://www.governance.ai/post/what-do-we-mean-when-we-talk-about-ai-democratisation.
70. Koelble, T.A., Lipuma, E.: Democratizing democracy: a postcolonial critique of conventional approaches to the 'measurement of democracy.' Democratization **15**, 1–28 (2008). https://doi.org/10.1080/13510340701768075
71. Fukuyama, F.: Political order and political decay: from the industrial revolution to the globalization of democracy. Macmillan (2014)
72. Mamdani, M.: Citizen and Subject : Contemporary Africa and the legacy of late colonialism /. Fountain Publishers, Kampala (1996)
73. Scott, J.C.: Seeing Like A State: How Certain Schemes to Improve the Human Condition Have Failed. Yale University Press (2020)
74. Shin, D.C.: On the third wave of democratization: a synthesis and evaluation of recent theory and research. World politics. **47**, 135–170 (1994)
75. Stokols, D., Hall, K.L., Taylor, B.K., Moser, R.P.: The science of team science. Am. J. Prev. Med. **35**, S77–S89 (2008). https://doi.org/10.1016/j.amepre.2008.05.002
76. Hinchliffe, S.: Technology, power, and space—the means and ends of geographies of technology. Environ. Plann. D Soc. Space. **14**, 659–682 (1996)
77. Miles, C.: The combine will tell the truth: on precision agriculture and algorithmic rationality. Big Data Soc. **6**(1), 2053951719849444 (2019). https://doi.org/10.1177/2053951719849444

(Im)possibilities in the Ethics of AI: Biometric Surveillance, Complicity, and Abolitionist Refusal in India and Beyond

Mallika G. Dharmaraj(✉)

University of Cambridge, Cambridge CB3 9HR, UK
nd513@cam.ac.uk

Abstract. In this paper, I address the question of "democratizing" AI by positing a discontent framework that unsettles the very possibilities inherent in "AI Ethics" as a project. I use the question of biometric surveillance in India as an attendant case study (with its centuries of Brahminical, colonial, and nationalist violence) but conclude with notes of relinquishment applicable broadly. To begin, I conduct a literature review of two key schools of AI Ethical thought—(1) a neoliberal fairness-centric first-wave (2) a staunchly sociotechnical second-wave of largely Black feminist criticism, which has sought to disrupt assumptions inherent in the former. Extending this critique, I then argue that the most ethical framework of AI in modern-day India seeks not to resolve, but to abolish and requires a politics of relinquishment from the dominant-caste Hindu (particularly those of us with ties to Silicon Valley's technological violence). Drawing inspiration from scholars who have keenly applied nuggets from Black feminist, abolitionist, and queer theory to the realm of the digital, this paper seeks to join a critical wave of scholarship that abandons the Western-neoliberal project of "AI Ethics" as such in favor of a reimagined site of complicity, refusal and hopelessness, guided by epistemologies from the margins—in short, this paper endeavors to discuss the (im)possibilities inherent in reconceiving an Ethics of AI.

Keywords: complicity · caste · Hindu nationalism · biometric surveillance · AI Ethics · India

1 Contested Sites: AI and Its Ethics

Despite its recent rise to popularity in the cultural, political, and social milieu of the twenty-first century, Artificial Intelligence (AI) has long stood as a disputed domain. For theorist Yarden Katz [1], author of *Artificial Whiteness*, AI itself is ontologically unsubstantiated as a term, embodying an "'unreal-yet-real'" characteristic—"unreal because it is nebulous, unstable, lacking in solid grounding, and continually redefined by powerful interests in ways that seem arbitrary…real in that it is part of concrete, destructive practices" (p. 167)—while Meredith Whittaker suggests that AI is intrinsically a "surveillance technology" [2]. Put simply, AI has always appeared as a refraction of cultural and political captures of it. As scholars Arvind Narayanan and Emily Bender have argued, this phenomenon has reached a peak in our current moment with fantastic, mythical discourses of "AI Hype" dominating media narratives [3, 4]. Even more, the chronology of AI, from its 1950s conception within the US military to the pessimistic 1980s "AI Winter" to what some enthusiasts now term our "eternal spring," has always been tightly embroiled within structures of power, hegemony, and resistance. In simpler terms, from its very origin, the history of AI has always also been a history of AI Ethics [1, p. 4].

As early as 1948, Western scholars of robotics were mathematically and philosophically contending with questions of what could be considered AI Ethics: roboticist Norbert Wiener and fiction writer Isaac Asimov stand as some of the first (Western) canonical academics to textually engage with issues of the ethical impact of AI technologies [5]. Simultaneously, many Indigenous philosophies of relationality, morality, and community that have existed for millennia around the world now apply agilely to the critical study of AI as well and demand recognition as precursive forms of AI Ethics. These include formulations from the working group Indigenous AI [6] to Amana Raquib et al.'s [7] Islamic objective (maqāṣid)-based ethics to AI to Sabelo Mhlambi's [8] Ubuntu conception of AI, and many other progressive critiques.

In this paper, I conduct an (inexhaustive) literature review of two key schools of AI Ethical thought—(1) a neoliberal first-wave, which draws from canonical Western philosophy and privileges the mathematical/technical, seeking to leverage the power of statistics to quantify, measure, and remediate problems in machine learning along criteria of fairness and accuracy (2) a staunchly *socio*technical second-wave of largely Black feminist criticism, which has sought to challenge and disrupt assumptions inherent in the first and link what we know as AI to projects of empire, statehood, capitalism, patriarchy, and hegemony. As a goal of these analyses, I hope to reroute the lingering question of "democratizing AI"—a debated term that has come to signify broadening the scope of participation in AI use, development, profits, and governance [9]—by visiting an abolitionist genealogy. That is, democratizing "AI," no matter how worthy in the abstract, still implies the existence of said "AI" and, with it, multiple attendant processes of social un/making. By contrast, the challenges posed by scholars critical of mainstream AI Ethics interrupt the premise of democratizing discourses by positing a discontent framework that unsettles the very possibilities inherent in "AI Ethics" as a project at all. I use modern-day India as an attendant case study, with its centuries of colonial, brahminical, and Hindu nationalist violence that empower a regime of biometric surveillance yet conclude with notes of relinquishment applicable broadly. In short, I argue that the most ethical framework of AI in modern-day India seeks not to resolve, but to abolish and requires a politics of relinquishment and refusal from the dominant-caste Hindu subject (particularly those of us with ties to Silicon Valley's technological violence). Drawing inspiration from the work of scholars who have keenly applied nuggets from abolitionist, Black feminist, and queer theory to the realm of the digital, this paper seeks to join a critical wave of scholarship that abandons the Western-neoliberal project of "AI Ethics" as such in favor of a reimagined site of resistance and even hopelessness (a la manmit singh [10]), guided by epistemologies from the margins. In short, this paper endeavors to discuss the (im)possibilities inherent in reconceiving an Ethics of AI.

2 Biometric Surveillance in India and Complicity: A Primer

In 2020, spurred by violently Hindu nationalist riots linked to Prime Minister Narendra Modi's xenophobic CAA/NRC legislation, the question of facial recognition technology (FRT) and CCTV cameras rose to the fore of the Indian political conversation. Recorded footage and spotty identification algorithms were used to make several targeted arrests (predominantly of Muslims) in the ensuing months. A recent Right to Information Application (RTI) launched by the Internet Freedom Foundation (IFF) discovered that, within the 2020 Delhi riots alone, FRT was used to identify more than 1900 individuals and arrest at least 137 individuals [11]. These findings are even more concerning given the recent rise of the National Automated Facial Recognition System (NAFRS) and imminent concerns about its integration with the world's largest biometric database, Aadhar [12]. Put simply, digital technologies like FRT now threaten to intensify an already-extant culture of state surveillance within an increasingly ethno-fascist India.

To make sense of these new techno-political imbrications, a sub-field of critical caste and technology scholars have begun to aptly historicize the project of surveillance. As sociologist Shivangi Narayan [13] reminds us, "there is little radically new about data-driven predictive policing systems" (p. 113). Rather, Big Tech has now merely reimagined what lawyers Ameya Bokil, Nikita Sonavane, et al. [14] term the "centuries-long project of predictive policing" in the subcontinent – analog surveillance technologies, caste/race-scientific eugenics, British criminal statutes, and brahminical mores have animated the biometric project in India since well before the advent of the modern computer. Incisive as they are, these interventions now leave a gaping open question for technologists and policymakers alike: what does a true Ethics of AI in India look like, when casteist and Islamophobic logics of policing and surveillance have been foundational to Indian society for centuries?

As we pause at these legacies of unimaginable violence, the question of complicity arises as a key operational concept. In the context of South Asia and its diasporas, complicity – across interlocking axes such as anti-Blackness, caste, and settler colonialism – has been brilliantly theorized in works by Shaista Patel & Dia Da Costa [15], manmit singh [10], and Shaista Patel, Moussa Ghaida, & Nishant Upadhyay [16], more broadly drawing from the instructive theory of scholars like Eve Tuck, Fiona Probyn-Rapsey, and Sara Ahmed. Meanwhile, in the specific realm of AI, Meredith Whittaker [17] and Corinne Cath & Os Keyes [18] have detailed the alarming corporate capture of critical AI work. However, little attention has been devoted to merging these traditions and tracing out an Ethics of AI that takes seriously questions of complicity, humility, and reflexivity as primary modes of engagement with the violence of (now)-digital surveillance.

Two years ago, I began the historical project of tracing the roots of digital and biometric surveillance in India with both curiosity and concern, in hopes of formulating a more humane Ethics of AI in India. Although I did not expect it, as I uncovered the brahminical support and Big Tech funding behind violently eugenicist ideologies, it became clear to me that I could not separate my own childhood, identities, and life experiences from this work. I was, and am, left reminded of the inescapability of my own complicity, as an incredibly wealthy/one-percent member of the Brahmin-savarna[1] non-Black Indian tech diaspora and a beneficiary of generational caste capital, exploitative logics of surveillance, and familial power in tech and venture capital. Rather than just sharing a mere litany of positionality markers, I write about these locations to think through the engagement of *complicity*. The ease with which I glossed over my own deep spiritual, material, and financial investments in the violence I sought out to study reflect not only my own failures to correct my learned "Brahminical ignorance, violence, and lack of basic education in ethics" [15, p. 18], but also, more broadly, an institutional unwillingness to approach questions of AI Ethics today with any meaningful attention to what feminist scholars have long termed "situatedness." Riffing off the work of Donna Haraway [19], "situated knowledges" are those embodied knowledges which grapple seriously with questions of complicity, reflexivity, and accountability at the forefront of any scholarly investigation, understanding that "unlocatable" work is accordingly "irresponsible" in a world where epistemic and material violence are intricately connected (p. 583). I am still sitting with extractive nature of projects like mine and the need to "consider the intimacy between privilege and the work we do, even in the work we do *on* privilege" [63] – where even stating a commitment to complicity, even writing a paper like this one, still so often becomes way of reinscribing violence and capital for dominant-caste people [10, p. 2]. Although rooted in my individual perspective, I share these reflections not merely as a self-reflexive exercise, but also as a means of beginning to theorize complicity, ethics, and abolition more broadly for the many of us who may occupy positions of privilege and power within critical technology discourses.

In short, inspired by my own irresolvably (un)ethical relations to the context of India but applicable to many contexts, I develop and proffer a reimagined Ethics of AI. That is, drawing a straight line between today's surveillant project of "AI" and the colonial/anti-Black/capitalist/brahminical world order that has operated on biometric logics for centuries, I use my research project in India to argue that any truly ethical "Ethics of AI" must take inescapable complicity as an invitation – starting and ending with deep reckonings around my/our own role in the violence foundational to our lives as scholars, thinkers, and people with any amount of capital.

I endeavor to pay special attention to citational ethics throughout this paper, foregrounding the work of Black, Indigenous, Dalit-Bahujan, and/or Muslim scholars. At the same time, in my writing, I wish to center *self*-critique in an effort to avoid what Shaista Patel and Dia Da Costa [15] importantly name as "the predatory citational practices of citing texts from racialized scholars on the margins only to criticize and make [my] analysis stand out as sanctimonious" (p. 2). Particularly in my engagement with Black feminist scholarship, I remain deeply cognizant of the often-predatory citation of Black feminism by non-Black scholars to earn "a valuable left credential" (p. 85), as theorized by Jennifer Nash [20]. I hold this contention close as I learn to intentionally hone a commitment to humility, care and respect in engaging life-giving Black feminist theory.

---

[1] Brahmin denotes the most powerful caste group in Indian society, while savarna is a term that connotes all members of (dominant-)caste society. I use the hyphenated term Brahmin-savarna to reflect myown maternal and paternal ancestry (my father being Tamil Brahmin, and my mother being Kaayasth – a dominant-caste in North India).

3 Philosophy and Mathematics: Classical Paradigms of Fairness

Firstly, over the last few decades, classical paradigms of AI Ethics have sought to harness the power of statistics to make corrective interventions into machine learning pipelines. In pursuit of developing algorithms that are aligned on axes of fairness and bias, ethicists of this tradition largely draw from classical mathematical and philosophical theory. Here, I review the hallmark contributions of this tradition. Crucially, as Harini Suresh and John Guttag [21] are quick to note, most all these paradigms fundamentally rest upon the assumption that "an underlying procedural or statistical notion of fairness can be mathematically defined and operationalized to create a fair system" (p. 3), a belief which I problematize below.

No doubt, paradigms of bias and fairness are methodologically informed by deep investments in the Western episteme. Renowned theorists Cynthia Dwork, Moritz Hardt, et al. [22] explicitly cite John Rawls, H. Peyton Young, and John Roemer as intellectual inspirations for their notions of individual- and group-fairness (p. 216). Meanwhile, several contemporaneous AI Ethicists subscribe to a school of thought known as long-termism, drawing from the movement for Effective Altruism (EA). Following the suit of Nick Bostrom [23], whose magnum opus explores the existential risks of "general superintelligence" (p. 277), this canon of philosophers subscribes to a mythical techno-utopian belief that (a homogenous) humanity's very continued survival is the greatest ethical challenge posed by contemporary AI technologies. Concerningly, philosopher Émile Torres [24] has helped highlight this movement's deep linkages to race science and liberal eugenics, citing Timnit Gebru's public critiques of long-termism.

Yet another school of optimistic data scientists has taken up the ill-defined term "social good" as a guiding principle of innovation. Information scientist Ben Green [25] states that these projects often nebulously set out to pursue a benchmark of "good" without any clear normative, material, or political analysis, investing in incrementalism and "reformist reforms" that often end up causing concrete harm. Green cleverly posits that, in this sense, social good and its investment in regulatory reformism (rather than structural change) analogously mirror the concept of a "greedy algorithm," any procedure which sets out to "make immediate improvements in the local vicinity of the status quo" but fails in more "complex search spaces" (p. 3). Similarly limited by neoliberal myopia, social good paradigms often seek to improve a small-scale, decontextualized problem identified by uninformed computer scientists, instead of leveraging data to center, collaborate, and resist in solidarity with communities in their long-term struggles.

The resulting fixations on bias and fairness are apparent in the literature. In 2019, Suresh & Guttag [21] conducted a birds-eye review of AI Ethics scholarship, highlighting various types of algorithmic fixes data scientists propose to make at various points in the machine learning pipeline: data-based, model-based, or post-hoc. Summarily, they enumerate five primary sources of "bias" in machine learning as well: historical bias (based on structural problems in society), representation bias (based on insufficiently diverse training sets), measurement bias (based on different proxy features across groups), aggregation bias (based on flawed assumptions of consistency across groups), and evaluation bias (based on an inaccurate benchmark that does not capture the underlying population well). Meanwhile, Hardt et al. [26] have pioneered many of the key frameworks for addressing algorithmic fairness in a way cognizant of "redundant encodings" (the existence of certain 'proxy' features which can serve as predictors for other sensitive attributes, thus making problematic any 'blind' approach that opts to simply ignore sensitive attributes). Foundationally, in 2011, Dwork et al. [22] formulated the Lipschitz condition of individual fairness for binary classifiers, guided by the concept that "any two individuals who are similar with respect to a particular task should be classified similarly" (p. 1). Another popular schema for fairness suggested by Dwork et al. is that of statistical parity, which works to ensure the maintenance of group fairness. Across these different paradigms, the overarching goal is to rigorously ensure equal likelihoods for different outcomes across subpopulations. While these aims sound promising in the abstract, given the kinds of structural injustices discussed above, the focus on subpopulation equity may be misdirected and even futile (i.e., what is gained by an FRT algorithm with equally-accurate classification likelihoods on Hindu and Muslim faces, when centuries-old brahminical-colonial forces, legal structures, and asymmetric concentrations of power have created the genocidal conditions for Muslims to be exclusively criminalized today?).

Some more progressive iterations of fairness research have attempted to reckon with social scientific theories. In 2019, Foulds et al. [27] proposed the metric of differential fairness (DF), informed by the theory of intersectionality as defined by Kimberlè Crenshaw. They seek to uphold five main criteria of intersectionality: the simultaneous regard of multiple sensitive attributes; protection of all possible intersections of said attributes; the continued guarantee of individual protected attributes; an eye towards marginalized/minority groups; and the broader correcting of structural disparity between societal groups. Intuitively, DF obeys the intersectional principle that outcomes will not vary regardless of one's (multiple) membership(s) in protected groups and thus presents itself as a mathematical way to enforce frameworks such as Crenshaw's. However, in response, philosopher Youjin Kong [28] has noted that Crenshaw's vision of intersectionality revolves around the intersection of oppressive sociopolitical *structures*, not the mere intersection of *attributes*. It cannot be so superficially quantified.

In the context of FRT specifically, AI Ethicist David Leslie [29] offers a summary of several popular strategies for bias mitigation among facial recognition developers today. A quick survey of these tactics makes clear that, like the frameworks posited above, they all seek to improve facial recognition's accuracy on marginalized/underrepresented groups, rather than questioning the political impulse behind surveillance technologies in the first place. Thus, classical paradigms in AI Ethics have focalized upon concepts of social good, long-termism, fairness and bias, with deep contributions from Western traditions (i.e., Enlightenment thinking, eugenics, and neoliberalism) to the epistemological underpinnings of this canon.

4 On Violence and Power: A Critical Black Feminist Second-Wave

However, in conjunction with these classical paradigms of fairness, recent decades have also borne witness to a more critical second wave of AI Ethical scholarship, predominantly led by Black feminist thinkers, that has sought to dislodge the very normative foundations of AI Ethics at all. Such scholars now challenge the Western white male as the humanistic base upon which most AI Ethical frameworks are built [30] (in the Indian context, it is crucial we extend perhaps further as well to the Hindu Brahmin male, given M.S.S. Pandian's reminder that "what looks like the unmarked modern is stealthily upper caste in its orientation" [31, p. 1738]). Applying ideology from abolitionist, feminist, and anti-caste movements to the study of technology, this wave of scholars has generated a rich and growing body of sociotechnical work that forces us to reckon with AI and its ensuing ethical (im)possibilities on a deeper, more radical level.

To begin, many thinkers have called into question the utility of "fairness" as an operational metric for AI Ethics, arising from a Science and Technology Studies critique that algorithms are not merely statistical instruments but cultural entities that animate specific social orders and worlds [32]. For one, building on previous scholarship that problematized fairness in the context of welfare allocation, Maximilian Kasy & Rediet Abebe [33] highlight three key drawbacks of fairness-centric approaches: their perpetuation of the fraught concept of "merit;" their concern with the narrow context of a single algorithm, blind to larger structures of power; and their ignorance of sub-group inequalities. The authors go on to suggest rigorous mathematical alternatives that prioritize political economy concerns, such as the distribution of power and inequality/causal impact. Similarly, Ben Green and Lily Hu [34] argue that the machine learning community's turn to fairness "as constituted by satisfaction of the statistical constraints is mistaken" and "mythic" (p. 2), while researcher Pratyusha Kalluri [35] evocatively warns, "Don't ask if artificial intelligence is good or fair, ask how it shifts power." In a keynote lecture, Ruha Benjamin [36] summed up this phenomenon as "techno-benevolence," or "tech developments that claim to address the bias of various sorts but may still manage to reproduce or deepen discrimination, in part because of the narrow way in which fairness is defined and operationalized." Frameworks such as Lipschitz normalization or equalized odds fail to address questions of context, power, and violence (as Green & Hu [34] paradigmatically demonstrate with the case study of fairness's inefficacy in COMPAS recidivism risk prediction). Approaches like statistical parity, on the other hand, have been faulted for falling into the liberal trap of flattening the hierarchy inherent to social attributes like race or gender in the first place [37]. Ultimately, many of these approaches to fairness epitomize a culture of what Jenny Davis, Apryl Williams, and Michael Yang dub as "algorithmic idealism – a meritocratic misconception of the world and a political ambivalence that this fallacy permits" [65].

Beyond just the topic of fairness, however, this wave of scholars has also launched a deeper epistemological critique. For instance, Khadijah Abdurahman [38] draws attention to the "hermeneutic injustice" of identity, jargon, and experience common in AI Ethical papers: while Fairness, Accountability, and Transparency (FAT*) conference papers spill pages of ink on the technical challenges of risk prediction, what goes unrecognized is "African American Vernacular English's (AAVE) rich ontological understanding—that never makes it to the conferences—but *been* understood the limitations of fairness grounded in identifying technical shortcomings of algorithmic decision making." From this place of critique, Alex Hanna et al. [37] proposed a groundbreaking critical race methodology for algorithmic fairness. They powerfully assert the need for questioning the project of automated classification itself as an act of structural violence, while also underscoring the necessity for disaggregation of racial data, treating race as a multidimensional social construct, centering the lived experiences of marginalized communities through non-ideal standpoint methodology, and focusing on the effects of *racism* (a system of power) rather than those of *race* (a fabricated identity). Similarly, researcher Lelia Hampton [39] advocates for an abolitionist, Black feminist reading of algorithmic oppression, powerfully stating: "algorithmic oppression cannot be overcome without integrating a Black feminist lens, a lens which empowers all peoples, a lá its stance that every form of oppression must be abolished" (p. 3). Revisiting Nash's warning [20] above, I wish to pause here to interrogate and interrupt my own learned urges to tokenize Black feminism as some progressive remedy that exists to solve all y/our problems. However, despite these relentless patterns of consumption by non-Black scholars, Hampton's words rightfully remind us of an inalienable truth. Black feminist ways of knowing make plain what has been praxis for marginalized communities resisting structural violence well before the development of digital technology—that "we can never have 'equality,' 'fairness,' or any other neoliberal buzzwords in AI until we abolish the white supremacist capitalist ableist ageist cisheteropatriarchy" [39 p. 4]. Researchers in surveillance studies have repeated such analyses, centering Blackness as a centuries-old locus of surveillance stretching from chattel slavery to the modern digital violence of biometric policing. Ruha Benjamin [40] argues that we must understand "anti-Black racism" not only as "a symptom or outcome, but a precondition for the fabrication of such technologies" (p. 44), while Simone Browne [41] explains that "Surveillance…is the fact of antiblackness" (p. 8).

Relatedly, a sea of scholars has also begun to marry decolonial theory from the Global South into AI Ethics. In 2016, mathematician Syed Mustafa Ali [42] introduced the scaffolding for "decolonial computing," aiming to center those at "the peripheries of the world system." Critiquing the previously popular framework of "postcolonial computing" for its Eurocentrism and class blindness, Ali's methodology focuses specifically on the geopolitical/biopolitical legacies of colonial epistemologies (through taxonomy, classification, categorization, etc.) in modern computational disciplines (p. 21). Philosopher Stephen Cave [43] contends that a cultural fetishization of "intelligence" itself, arising from colonial projects of eugenics and pseudoscience, motivates many AI discourses today. Legal scholar Chinmayi Arun [44] summarizes ways in which AI social orders may particularly impact vulnerable populations in the Global South and thus demand understanding within "different models of exploitation" (p. 595). Accordingly, Data & Society researchers Sareeta Amrute, Ranjit Singh, and Rigoberto Lara Guzmán [45] compiled an excellent primer of sources on AI from the "Majority World"—as not just as a geographical region, but a methodological framework to "understand, analyze, and build developmental, postcolonial, and decolonial computing practices" (p. 6). Across forty-four pages, they explore topics ranging from anti-caste culture to Afro-Modernities to surveillance and extraction. Finally, the turn towards decolonial AI has been most notably summarized by Shakir Mohamed, Marie-Therese Png, and William Isaac [46] in a landmark 2020 study. Engaging various facets of decolonial theory, Mohamed et al. coin the term "algorithmic coloniality" to capture the manifold ways in which data enables labor exploitation, spatial divides between metropole/periphery, and digital dispossession. They conclude by suggesting intercultural dialogue, reverse power tutelage, and affective communities of solidarity as paths forward in a critical technoscience approach.

Without a doubt, these incredible works provide rich and instructive theories on global power that I have often shared, revisited, and appreciated. However, in the pursuit of holding myself accountable and interrupting the incessant desire to locate (my) comfort within complicity, I am also interested in the ways that even these decolonial departures from normative AI Ethics deserve scrutiny. For one, there must be a sole, specific, incommensurable focus on Indigenous repatriation in all projects of decolonization, as Eve Tuck and K. Wayne Yang [47] remind us: "[w]hen metaphor invades decolonization, it kills the very possibility of decolonization…Decolonize (a verb) and decolonization (a noun) cannot easily be grafted onto pre-existing discourses/frameworks, even if they are critical, even if they are anti-racist, even if they are justice frameworks" (p. 3, 21). By the same token, Hampton [39] asks: "How do we decolonize AI if the world is not even decolonized?" (p. 4). Meanwhile, anti-caste scholars in the Indian political tradition have long contested decoloniality as a worthwhile lens du jour for caste liberation at all, given the deep preexistence of brahminical violence for millennia prior to the colonial encounter and the ways that a "decolonizing" agenda has been co-opted by the Hindu right [48].

5 Abolition, Refusal and Hopelessness: On Contemporary FRT in India

I now turn to the state of contemporary FRT in India, with its centuries-long informant histories of brahminism and British colonialism. Clearly, the utmost intention and care is required in the choice of framing even critical, radical, and justice-oriented interventions into AI Ethics. In toto, the aforementioned body of Black feminist work gestures towards the fact that projects of fairness, equality, and ethics are often used to sanitize corporate/state-sanctioned harm—in a move that scholars like Timnit Gebru [49] and Elettra Bietti [50] have called "ethics washing." By extension, then, it is worth asking about the very (im)possibility of writing any "Ethics of AI" at all, so long as AI is materially affixed to systems like white supremacy, capitalism, and brahminism.

In Katz' [1] eyes, this is evident, as he argues that the fundamental logic of American AI Ethics itself is one of neoliberal acceptance, as also noted by Rodrigo Ochigame [66]. In the eyes of AI Ethicists, "[t]he way to cope is to expand: to develop more codes of 'ethics' and mechanisms of accountability" (p. 4). In lieu, he calls for a "generative refusal" that acts from a place of intellectual traitorship, appreciating the irredeemable legacies of structural violence around AI institutions and elevating coalitional movement work over ethical pontification. Such a stance echoes Seeta Peña Gangadharan's [64] elaboration of our "right of refusal" to disobey and deny technologies of control. No doubt, refusal occupies a heralded space in many schools of radical thought. In *The Great Refusal: Herbert Marcuse and Contemporary Social Movements*, a variety of scholars and activists [51] subversively subscribe to an "ontology of negativity…in antagonism and refusal," which "oppose[s] the false assurances of reconciliation" and "'refuse[s] to play the game'" (p. 55–65, 260)· While Katz' reasoning remains perhaps limitingly bounded to the conditions of power that follow from whiteness alone, singh

[10] transposes a similar politic to the nuanced context of movement work within Indian hierarchy, inspired by trans Sikhi subjectivities—what they term a "pedagogy of hopelessness," which asks us "to interrogate our desire to map and discover the landscapes of the imagination, of the 'not yet,' in ways that mirror scripts of conquest already too familiar" (p. 11). For singh, there is a particular exigency for us as "non-Black, non-Indigenous, non-Muslim, dominating caste" peoples to adopt hopelessness, with all its hues of grief and despair, relinquishing "our desire for an 'otherwise'" and instead honing "our capacity to meditate on suffering" (p. 11).

I thus seek here to both join and advance the latter wave of critical research by pairing the call for a "generative refusal" in American AI with singh's "pedagogy of hopelessness." In substance, this section positions two key points to the fore: firstly, the relevance of non-Western/Indian-specific contextual understandings in discourses of AI Ethics, and secondly, an abolitionist (rather than reformist) understanding of state surveillance and policing. Ultimately, I in fact can and do not claim to offer any reconceived Ethics of AI in India, disrupting the oppressive, dominant-caste Hindu, and neoliberal gravitation towards expedient remediation and reconciliation in the face of genocide. Rather, I heed singh's call to abandon the reconciliatory register of "hope" and return instead to "what *remains*," asking to center caste- and religious-marginalized grief, vision, and resistance in the fight towards abolition of the now-computerized police state (p. 11).

Firstly, I argue that there is a deep incommensurability of Western AI Ethical frameworks with the nuanced structural realities of brahminism and colonialism in India. These incongruencies alone necessitate a stance of staunch refusal. In 2021, Google researcher Nithya Sambasivan et al. [52] released a paper that originally argued this very point. Sambasivan et al. emphasize the Western orientation in fairness research, asking the questions: Which fairness researchers are seriously thinking with caste-marginalized Dalit or Adivasi communities in India? Why do the writings of John Rawls command so much intellectual capital among the fairness community, as opposed to those of B.R. Ambedkar or Indigenous restorative justice practitioners? How do US-centric legal framings (like "disparate impact") translate into Global South contexts? In India, specifically, context-based issues include glaring gaps in datasets excluding marginalized caste/religious communities, misreporting of identity attributes, unique sub-group proxies, overfitting to privileged identities, and blind cultural euphoria around AI. Aside from these keen observations, Sambasivan et al. also comment on the particularities of power and surveillance in India. From interviewing legal experts and auditors, it became clear that "Dalit and Muslim bodies were used as test subjects for AI surveillance" in places of work, echoing centuries-long traditions of surveillance of the working poor by dominant-caste employers e.g., think here of "'the house cleaner who is constantly suspected of stealing dried fruits or jewellery'" (p. 7).

Ultimately, Sambasivan et al.'s propositions are crucial to unsettling the assumptions behind AI Ethics in India, as they correctly state the need to foreground caste, gender, national politics, and hierarchy. Like Tuck, Yang, and Hampton, their position favors Indian contextual particularity over the often nebulous and problematic lens of decoloniality. They assert that "[t]o Dr. B. R. Ambedkar and Periyar E. V. Ramasamy, colonialism predates the British era, and decolonisation is a continuum" (p. 4). Dwivedi et al. [53] have gone even further in importantly reminding us that in fact, "colonialism provided [the oppressed castes of the subcontinent] emancipatory conditions," as they endeavored to "seize the fortunes offered to them." These observations on the limited theoretical utility of decoloniality coincide with Dia Da Costa's [54] "multiple colonialisms" analysis, which grapples with "intersecting structures of precolonial violence (e.g., caste) and casteist coloniality in one time and space (e.g., so-called postcolonial India) to consider its relational constitution with conventionally-recognizable colonialisms (e.g. British colonization of India)" (p. 505). Camille Acey et al. [55] deftly extend this analysis to the realm of the digital as well, forwarding "debrahminizing as decolonizing on Wikipedia" as objectives for caste-marginalized communities (p. 3), and Dr. Murali Shanmugavelan's [56] noteworthy "Critical Caste and Technology Studies" syllabus has called recent attention to the myriad junctures of casteism and digital life in India.

While these excellent scholarly positions necessarily complicate the broad preconditions of AI Ethics in an Indian context, there remains another unresolved contention around FRT specifically: the question of abolition versus reform. That is, bearing in mind Hampton's [39] provocation that algorithmic fairness researchers often "do not consider that some of these technologies should not even exist" (p. 6), I posit that the only ethical stance towards FRT in India is an abolitionist one. For, if we accept that Indian FRT arose (not tangentially, but primarily) as a direct digital continuation of brahminical-colonial investments in criminalizing the marginalized, there can be no "more ethical" iteration. Instead, precisely because of historical and present dominant-caste complicity in Indian state surveillance, the onus is on us as savarna elites and diasporic Hindus to reject the impulse toward any reordering of AI Ethics in this moment. A truly ethical struggle, informed by refusal and hopelessness, is one that rejects plans of reform or democratization and moves towards the utter annihilation of the casteist, Islamophobic, and criminalizing ideologies in our cultures which have made such a system possible in the first place.

The conversation on American FRT has borne witness to analogous reckonings. In 2018, Joy Buolamwini and Timnit Gebru [57] released the controversial "Gender Shades" project, which called for greater transparency, accountability, and inclusion along lines of skin color and gender in facial recognition algorithms. In its wake, various abolitionist voices noted that, regardless of intention, such interventions epitomize carceral thinking and foster the growth of the American police state's panoptic eye for better, fairer, and more accurate policing of Black faces in the name of liberal reform (similar logics have been deployed in service of the Israeli surveillance state as well). Media theorist Nabil Hassein [58] cites Black liberationist theory to suggest refusing "the development or deployment of technology which makes it easier for the state to recognize and surveil members of my community," including FRT and all other future-looking forms of biometric surveillance as well. Meanwhile, Benjamin [40] asks: "[w]hile inclusion and accuracy are worthy goals in the abstract, given the encoding of long- standing racism in discriminatory design, what does it mean to be included and hence more accurately identifiable, in an unjust set of social relations?" She concludes that "[i]nclusion in this context is more akin to possession." (p. 124). Critical knowledge of this sort has long existed deeply among African-American communities: Abudrahman [38] writes, "If you come to the hood...you would immediately be contested by people whose concern with predictive policing has nothing to do with how protected classes within algorithms are generated—but who viscerally reject the notion human beings should be placed in cages in the first place."

These abolitionist critiques adapt well to the Indian context, where there clearly exists a strong history of police violence invested in the renewal of hegemony and biopolitical control of the marginalized vis-a-vis surveillance technologies. We must understand surveillance as foundational for the caste and religious order, from the codes of Vedic elites to the writings of colonial administrators to the birth of the CCTV camera. Such historical convergences make one point staunchly apparent: contemporary Indian FRT is *not* a case of a 'neutral'/'good' technology in development that happens to generate unfortunately harmful effects in deployment. It is not even a case of a majorly 'biased' technology. Rather, the whole state-sanctioned, techno-capitalist life-cycle of FRT, like any past or future regime of biometric surveillance in India, is entirely constituted by and constitutive of a particular form of brahminical politics – one that is predicated on the subjugation, exploitation, and dispossession of those deemed criminal by birth. It is particularly urgent for us to adopt an ethic of uncompromising abolition and nothing less due to the ravenous cycle of Hindu nationalist co-optation. Given Modi's endlessly vicious propaganda machine [58], even radical, anti-caste formulations of AI Ethics are at danger of being consumed, alongside any other seemingly progressive Indian AI agenda, to the ends of reifying a liberalized (Hindu) Indian nation-state. Wholesale refusal remains the only possible ethic.

6 Conclusion: Inhabiting Deathworlds

In sum, learning from the wisdom of Black feminist and abolitionist criticism, I argue that discourses on AI and ethics in India must abandon classical philosophical and quantifiable paradigms of fairness, instead turning to a historically-informed perspective that centers power, structural violence, and liberation. Such a view must transcend Western Anglophone as well as Hindu brahminical normativities to center the systems of caste, gender, and religion while committing to a rigorous abolitionism that understands the digital surveillance cycle as an indivisible part and parcel of empire, statehood, and Hindu nationalism. I concentrate on complicity as a meditative starting point for the refinement of any such politic.

In practice, this means adopting a pedagogy of refusal and hopelessness as "non-Black, non-Indigenous, non-Muslim, dominating caste" people. In a time of genocide, we must first and foremost attend to our locations and obligations, recognizing that any truly ethical digital future in India can only stem from the ways of knowing, living, and resisting that have been central to the life of religious/caste-marginalized communities for centuries immemorial. Per Gopal Guru's [59] work, the trajectory of such work is constantly endangered by brahminical epistemic violence:

> As 50 years' experience shows, social science practice has harboured a cultural hierarchy dividing it into the vast, inferior mass of academics who pursue empirical social science and the privileged few who are considered the theoretical pundits

with reflective capacity which makes them intellectually superior to the former. To use a more familiar analogy, Indian social science represents a pernicious divide between theoretical brahmins and empirical shudras.

With an eye towards continued the brahminical intellectual ravaging and dehumanization of caste- and religiously-marginalized communities (see *Hatred in the Belly* [60]), Guru's words thus serve as an acute reminder that the dominant-caste elite's role is not to theorize liberation on the lives of those violated. Borrowing from the words of Shaista Patel [61], I conclude: as a savarna person inheriting the legacies of caste capital, British colonial surveillance, and diasporic Big Tech wealth, "I am not in an ethical place" to ask what lies "beyond" or "otherwise" in these digital futures [10]. My/our question cannot be what is possible in Indian AI Ethics. Our project cannot be to regulate, reform, reconceive, or democratize, as democratizing often merely means pluralizing access to violence.

Given the long histories of our complicity in state surveillance, I urge us instead to turn to death and impossibility as productive sites for the digital (bearing in mind Heather Love's analogous proposition that "death and impossibility" have always been homes for the queer subject within the violent cisheteropatriarchy) [50, p. 127]. As singh [10] writes, our principal ethical imperative as "non-Black, non-Indigenous, non-Muslim, dominating caste" is to "invest in our own death that is necessitated by the project to end the world as we know it" (p. 6). I extend their position by arguing that the world, "as we know it" today, leaks into the digital, so the counter-project of abolition must thus incorporate it too. In substance, this must minimally mean turning inwards to our barbaric cultures of caste criminality, colonial collusion, and Hindu statehood that have given life to regimes of digital surveillance, while also committing to sustained reparation of the wealth amassed from caste exploitation and slavery. In sum, I cannot provide a substantiated Ethics of AI in India, because to do so would only serve to offer us a place of reinscribed innocence and comfort in the face of continued dispossession and ethnic cleansing. To do so would be to further the imploding, accelerating project of attempting to repair the irreparable, as I pay heed to Logic(s) magazine's [62] editorial provocation: "Counter to Facebook's mantra, we must *move slow and heal things.*"

Rather, I ask us to turn our gaze to the possibility of impossibility, embracing a radically different pace than the "urgent tempo of crisis" – one of disquieting pause, slowness, and intention [62]. By turning to the contradiction and incommensurability inherent in any project of "Ethical AI" under a casteist, colonial, and capitalist world order, we in fact open up what I will call *death*-worlds of possibility which leave us with "no innocence and no refuge" [61, p. 229]. If we are to take seriously Black feminist epistemologies, it is worth thinking with Jennifer Nash's [20] words: "What if we imagined Black feminist ethical and political struggle not as providing the answer…but as demanding that we sit in the murky territory of not-yet-knowing?" (p. 87). Doing the disquieting work of probing our role in mass violence and working towards the investment in our own death asks us to put away our curiosities about the who, what, and how of digitally-"decolonized, debrahminized, abolitionist worlds" and commit to "not-yet-knowing"—a practice I see as central to my own (inherently partial) auto-theory. It asks us to adopt instead a continued rigorous ethical engagement with the millennia of histories and complicities that have allowed a regime like India's to capitalize on FRT in the way it has today. In short, if classical AI Ethics research is meant to breathe into AI new lifeworlds that represent greater "fairness" or "equality," then Black feminist, abolitionist thought suggests the inhabiting of a deathworld instead, that which understands hopelessness as instructive, refusal as requisite, and abolition as urgent. I wish to end with a fruitful and appropriate call to action and relinquishment—a practice of "learning and…deep listening"—that respects and celebrates Dalit, Bahujan, Muslim, Indigenous, and Black feminist agency, brilliance, and dynamism in the ever-evolving struggle for unsurveilled futurities [10].

Acknowledgments. This work is entirely a labor of community, and I would be remiss to take sole credit. I would like to thank Vikrant Dadawala and Michael Smith for their indelible mentorship as advisors to my undergraduate thesis (from which this paper was born). As well, I thank Amy Gaeta, manmit singh, Kerry McInerney, Maya Indira Ganesh, and Allanah Rolph for their detailed suggestions on this submission. Finally, I am indebted to the endless support from members of my chosen family Alexis Queen, Eden Fesseha, Bhargavi Garimella, and countless others, as well as to the organizing spaces that I have been grateful to share and grow my ethics within.

References

1. Katz, Y.: Artificial Whiteness: Politics and Ideology in Artificial Intelligence. New York, NY (2020)
2. Coldewey, D.: Signal's Meredith Whittaker: AI Is Fundamentally 'A Surveillance Technology (2023).' https://techcrunch.com/2023/09/25/signals-meredith-whittaker-ai-is-fundamentally-a-surveillance-technology/. Accessed 9 Oct 2023
3. Bender, E.: On NYT Magazine on AI: Resist the Urge to Be Impressed (2022). https://medium.com/@emilymenonbender/on-nyt-magazine-on-ai-resist-the-urge-to-be-impressed-3d92fd9a0edd. Accessed 9 Oct 2023
4. Zuckerman, E. 63: See Through AI Hype with Arvind Narayanan (2022)
5. Borenstein, J., Grodzinsky, F., Howard, A., Miller, K., Wolf, M.: AI ethics: a long history and a recent burst of attention. Computer **51**(1), 96–102 (2021)
6. Indigenous Protocol and Artificial Intelligence Working Group, https://www.indigenous-ai.net/. Accessed 9 Oct 2023
7. Raquib, A., Channa, B., Zubair, T., et al.: Islamic virtue-based ethics for artificial intelligence. Discov. Artif. Intell. **2**, 11 (2022)
8. Mhlambi, S.: From rationality to relationality: ubuntu as an ethical and human rights framework for artificial intelligence governance. In: Carr Center Discussion Paper Series (2020)
9. Seger, E., Ovadya, A., Siddharth, D., et al.: Democratising AI: multiple meanings, goals, and methods. In: AAAI/ACM Conference on AI, Ethics, and Society (AIES '23), pp. 715–722. Montréal, QC, Canada (2023)
10. singh, m.: The Ally Must Die: A Trans Sikh Politics of Death and Unbodiment. Master's Thesis. San Francisco State University (2023)
11. Face Recognition Software Used in 137 of 1,800 Arrests in Northeast Delhi Riots, Says Police. https://indianexpress.com/article/cities/delhi/delhi-riots-police-cctv-7196291/, Accessed 9 Oct 2023
12. Doddahatti, B., Bokil, A.: Under the Thumb: A New Legislation Expands the Government's Surveillance Powers. The Caravan (2022)

13. Narayan, S.: Guilty until proven guilty: policing caste through preventive policing registers in india. J. Extreme Anthropol. **5**(1), 111–129 (2021)
14. Bokil, A., Khare, A., Sonavane, N., et al.: Settled Habits, New Tricks: Casteist Policing Meets Big Tech in India. In Longreads (2021)
15. Patel, S., Da Costa, D.: 'We cannot write about complicity together': limits of cross-caste collaborations in western academy. Engaged Sch. J. Community-Engaged Res. Teach. Learn. **8**(2), 1–27 (2022)
16. Patel, S., Moussa, G., Upadhyay, N.: Guest Editorial: Complicities, Connections, and Struggles: Critical Transnational Feminist Analysis of Settler Colonialism. Feral Feminisms (2015)
17. Whittaker, M.: The Steep Cost of Capture. Interactions Magazine (2022)
18. Cath, C., Keyes, O.: Your thoughts for a penny? capital, complicity, and AI ethics. In: Phan, T. (ed.). Economies of Virtue – The Circulation of 'Ethics' in AI, pp. 24–38. Institute of Network Cultures, Amsterdam, The Netherlands (2022)
19. Haraway, D.: Situated knowledges: the science question in feminism and the privilege of partial perspective. Fem. Stud. **14**(3), 575–599 (1998)
20. Nash, J.: Citational desires: on black feminism's institutional longings. Diacritics **48**(3), 76–91 (2020)
21. Suresh, H., Guttag, J.: A Framework for Understanding Unintended Consequences of Machine Learning. arXiv:1901.10002v3 (2019)
22. Dwork, C., Hardt, M., Pitassi, T., et al.: Fairness through awareness. In: ITCS '12: Proceedings of the 3rd Innovations in Theoretical Computer Science Conference, pp. 214–226 (2012)
23. Bostrom, N.: Ethical issues in advanced artificial intelligence. In: Schneider, S. (ed.) Science Fiction and Philosophy: From Time Travel to Superintelligence, pp. 277–284. Wiley-Blackwell (2009)
24. Torres, E.: Nick Bostrom, Longtermism, and the Eternal Return of Eugenics (2023). https://www.truthdig.com/dig/nick-bostrom-longtermism-and-the-eternal-return-of-eugenics/. Accessed 9 Oct 2023
25. Green, B.: Good isn't good enough. In: AI for Social Good Workshop at NeurIPS, Vancouver (2019)
26. Hardt, M., Price, E., Srebro, N.: Equality of opportunity in supervised learning. In: 30th Conference on Neural Information Processing Systems, pp. 3323–3331. Barcelona, Spain (2016)
27. Foulds, J., Islam, R., Keya K., et al.: An intersectional definition of fairness. In: 2020 IEEE 36th International Conference on Data Engineering (2020)
28. Kong, Y.: Are 'Intersectionally Fair' AI algorithms really fair to women of color? A philosophical analysis. In: FAccT '22: Proceedings of the 2022 ACM Conference on Fairness, Accountability, and Transparency, pp. 485–494. (2022)
29. Leslie, D.: Understanding Bias in Facial Recognition Technologies: An Explainer. The Alan Turing Institute (2020)
30. AI Decolonial Manyfesto. https://manyfesto.ai/. Accessed 9 Oct 2023
31. Pandian, M.S.S.: One step outside modernity: caste, identity politics and public sphere. Econ. Pol. Wkly **37**(18), 1735–1741 (2002)
32. Seaver, N.: Algorithms as culture: some tactics for the ethnography of algorithmic systems. Big Data Soc. **4**(2), 2053951717738104 (2017)
33. Kasy, M., Abebe, R.: Fairness, equality, and power in algorithmic decision-making. In: FAccT '21, pp 576–586. Virtual Event, Canada (2021)
34. Green, B., Hu, L.: The myth in the methodology: towards a recontextualization of fairness in machine learning. In: Machine Learning: The Debates Workshop at the 35th International Conference on Machine Learning. Stockholm, Sweden (2018)

35. Kalluri, P.: Don't ask if artificial intelligence is good or fair, ask how it shifts power. Nature **583**(7815), 169–169 (2020)
36. Benjamin, R.: Race to the future? reimagining the default settings of technology and society. National Center for Women & Information Technology Conversations for Change, NCWIT (2020)
37. Hanna, A., Denton, E., Smart. A, et al.: Towards a critical race methodology in algorithmic fairness. In: FAT* '20: Proceedings of the 2020 Conference on Fairness, Accountability, and Transparency, pp. 501–512 (2020)
38. Abdurahman, J.K.: FAT* Be Wilin': A Response to Racial Categories of Machine Learning by Sebastian Benthall and Bruce Haynes. https://upfromthecracks.medium.com/fat-be-wilin-deb56bf92539. Accessed 9 Oct 2023
39. Hampton, L.: Black feminist musings on algorithmic oppression. In: FAccT '21: Proceedings of the 2021 ACM Conference on Fairness, Accountability, and Transparency. Virtual Event, Canada (2021)
40. Benjamin, R.: Race After Technology: Abolitionist Tools for the New Jim Code. Polity, Cambridge, UK (2019)
41. , S.: Dark Matters: On the Surveillance of Blackness. Duke UP, Durham, North Carolina, US (2015)
42. Ali, S.: A Brief Introduction to Decolonial Computing. XRDS Crossroads The ACM Mag. Students **22**(4), 16–21 (2016)
43. Cave, S.: The problem with intelligence: its value-laden history and the future of AI. In: AIES '20: Proceedings of the AAAI/ACM Conference on AI, Ethics, and Society, pp. 29–35. New York, NY, US (2020)
44. Arun, C.: AI and the global south: designing for other worlds. In: Dubber, M., Pasquale, F., Das, S. (eds.) The Oxford Handbook of Ethics of AI, pp. 588–606 (2020)
45. Amrute, S., Singh, R., and Guzmàn, R.: A primer on AI in/from the majority world: an empirical site and a standpoint. Data Soc. (2022)
46. Mohamed, S., Png, M., Isaac, W.: Decolonial AI: decolonial theory as sociotechnical foresight in artificial intelligence. Philos. Technol. **33**, 659–684 (2020)
47. Tuck, E., Yang, K.W.: Decolonization is not a metaphor. Decolonization Indigeneity Educ. Soc. **1**(1), 1–40 (2012)
48. Mani, B.R.: Beyond the Fairy Tales of India (2015). https://countercurrents.org/mani070415.htm. Accessed 9 Oct 2023
49. Gebru, T.: "This is what is called ethics washing (2021)." https://twitter.com/timnitGebru/status/1394717504119971844. Accessed 9 Oct 2023
50. Bietti, E.: From ethics washing to ethics bashing: a view on tech ethics from within moral philosophy. In: FAT* '20: Proceedings of the 2020 Conference on Fairness, Accountability, and Transparency, pp. 210–219. Barcelona, Spain (2020)
51. Lamas, A., Wolfson, T., Funke, P. (eds.): The Great Refusal: Herbert Marcuse and Contemporary Social Movements. Temple UP, Philadelphia, Pennsylvania, US (2017)
52. Sambasivan, N., Arnesen, E., Hutchinson, B., et al.: Re-imagining algorithmic fairness in india and beyond. In: FAccT '21: Proceedings of the 2021 ACM Conference on Fairness, Accountability, and Transparency, pp. 315–328. Virtual Event, Canada (2021)
53. Dwivedi, D., Mohan, S., Reghu, J.: The Hindu Hoax: How Upper Castes Invented a Hindu Majority. The Caravan (2020)
54. Da Costa, D.: Eating heritage: caste, colonialism, and the contestation of adivasi creativity. Cult. Stud. **33**(3), 343–369 (2019)
55. Acey, C., Bouterse, S., Ghoshal, S., et al.: Decolonizing the internet by decolonizing ourselves: challenging epistemic injustice through feminist practice. Global Perspect. **2**(1) (2021)
56. Shanmugavelan, M.: Syllabus (2022). https://criticalcastetechstudies.net/. Accessed 9 Oct 2023

57. Buolamwini, J., Gebru, T.: Gender shades: intersectional accuracy disparities in commercial gender classification. In: Proceedings of Machine Learning Research vol. 81, pp. 1–15 (2018)
58. Hassein, N.: Against Black Inclusion in Facial Recognition (2017). https://digitaltalkingdrum.com/2017/08/15/against-black-inclusion-in-facial-recognition/. Accessed 9 Oct 2023
59. Shahanem G.: Appropriating Ambedkar: Why the BJP is on Strong Ground in its Battle to Co-opt the Hindutva Critic (2018). https://scroll.in/article/875183/appropriating-ambedkar-why-the-bjp-is-on-strong-ground-in-its-battle-to-co-opt-the-hindutva-critic. Accessed 9 Oct 2023
60. Gopal, G.: How egalitarian are the social sciences in India? Econ. Pol. Wkly **37**(50), 5003–5009 (2002)
61. Ambedkar Age Collective: Hatred in the Belly: Politics behind the Appropriation of Dr Ambedkar's Writings. Shared Mirror, Hyderabad, India (2015)
62. Patel, S.: Talking complicity, breathing coloniality: interrogating settler-centric pedagogy of teaching about white settler colonialism. J. Curriculum Pedagogy **19**(3), 211–230 (2022)
63. The Editors. The Rain. Logic(s) (2023)
64. Ahmed, S.: Declarations of whiteness: the non-performativity of anti-racism. Borderlands **3**(2) (2004)
65. Gangadharan, S.P.: Technologies of Control and Our Right of Refusal. In: YouTube, https://www.youtube.com/watch?v=XPNMSUA2zxQ. Accessed 10 Oct 2023
66. Davis, J., Williams, A., Yang, M.: Algorithmic Reparation. Big Data Soc. **8**(2), 20539517211044808 (2021)
67. Ochigame, R.: The Invention of "Ethical AI (2019)." https://theintercept.com/2019/12/20/mit-ethical-ai-artificial-intelligence/. Accessed 23 Oct 2023

Research Methods of the Impact of AI on Elections – Systematic Review

Maria Lipińska(✉)

University of Warsaw, Warsaw, Poland
lipinska.maria123@gmail.com

Abstract. The paper provides an overview of the research methods used for examining AI's influence on elections and political processes. In the last few years, a democracy crisis has been observed in many countries in parallel with dynamic AI progress. The author conducted a systematic literature review using the PRISMA method on 18 papers published in the years 2019–2023 from Scopus. The subject of AI and elections is often presented in the context of fake news. There is an alarming research gap in conducting empirical studies on the impact of artificial intelligence on elections. This paper establishes a methodological foundation for further academic inquiry.

Keywords: AI · Elections · Voting

1 Introduction

Witnessing a global race of big tech companies to provide the best artificial intelligence, global institutions and governments should reconsider the impacts of technology on society and politics (Bareis and Katzenbach 2022; Schuett 2022). As previous research (P et al. 2023) rightly notes, AI can be implemented in five areas of elections: voter list maintenance and de-duplication, determining polling booth locations, vulnerability-based polling booth protection, voter authentication, and video monitoring of electoral fraud. The idea and practice of democracy can be severely affected by AI (Jungherr 2023). Moreover, during the last few years, a democracy crisis has been observed in many countries (Farkas and Schou 2020). We are facing difficulties connected with disinformation spread online, deep-fakes, algorithmic surveillance, and microtargeted political campaigns. Thus, solutions for empowering the people in democracy should be answered, as should the potential impact of AI on elections be discussed. In this paper, the author addresses crucial research questions: What is the impact of the election process on society? What are the most accurate research methods for examining AI's influence on elections?

Each area of science has difficulties associated with the definitional challenges of AI. This article adopts the definition from the OECD, as it compromises human and algorithmic roles. *An AI system is a machine-based system that can, for a given set of human-defined objectives, make predictions, recommendations, or decisions influencing real or virtual environments. AI systems are designed to operate with varying levels of autonomy* (Bedse et al. 2022).

Considering previous research on the impact of AI on elections, it can be observed that it is not the leading subject in the field of social sciences. To assess the topic's popularity in the field, the author analyzed the research scope (Reis et al., 2019) and the number of articles published each year in the Scopus database. After this phase, appropriate keywords were identified. The decision to focus on keywords related to "AI" or "artificial intelligence" in conjunction with "election(s)" or "voting" aims to address the research questions. The selection of the database was considered appropriate because it covers most of the IEEE and ACM journals Fig. 1.

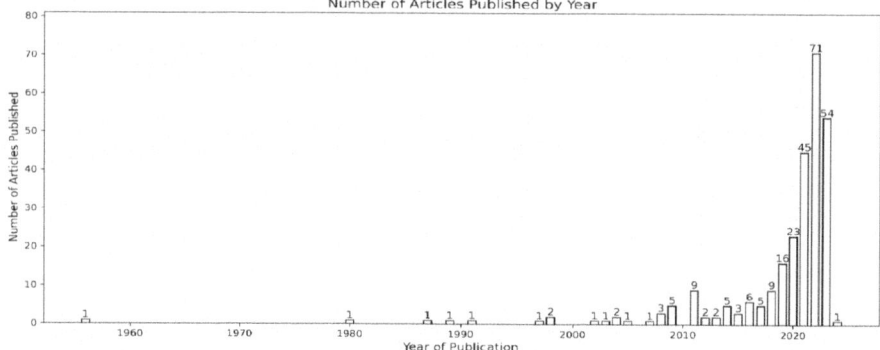

Fig. 1. Histogram of article publication dates, ABS - KEY("AI" OR "artificial intelligence" & "election*" OR "voting") AND (LIMIT - TO(DOCTYPE: ARTICLE)) AND (LIMIT – TO(LANGUAGE: ENGLISH). Source: Scopus.

Since 2019, there has been a notable increase in research on the impact of AI on elections. Researchers admit that far too little research is carried out in this area (Jungherr, 2023). Considering these factors, there is a research gap in examining the relationships between AI and its impact on elections. This paper intends to give a systematic overview of the research methodologies employed and showcase the main conclusions from articles published between 2019 and 2023 that are connected with the topic of artificial intelligence and elections. This timeframe was chosen for the analysis due to the largest peak in the number of publications. To examine this subject, the analysis consists of three areas: 1. Definition of AI, 2. Methods used to investigate the relationship between AI and elections, 3. The main conclusions drawn from the article.

2 Methods

In order to examine the research process in papers connected with AI's influence on elections, a systematic literature review was conducted in accordance with the PRISMA statement (Page et al. 2021). This methodology was applied due to the presentation of the methods of research on the topic of AI and elections. The analysis included coding categories (Mayring 2014) such as AI definition, methods, techniques of research including tools and number details (Reis et al., 2019), AI influence highlighted in the paper, and main conclusions. As it was mentioned before, papers were chosen from the Scopus database and published in the years 2019–2023 due to the popularity of the topic. The analysis was conducted at 15.10.2023. At the third stage of research, there were 41 articles listed. After applying manual inclusion-exclusion criteria based on the main research question and aim of the article, a list of 18 articles was compiled for the final analysis. The excluded articles were connected to the subjects of art, history, education, security and language.

Figure 2 presents three stages of research. The scheme (Fig. 2) is based on the PRISMA statement (Rethlefsen et al. 2021) and workflows presented in other papers using this methodology (Reis et al. 2019; Sharma et al. 2020; Secinaro et al. 2021, Reis et al., 2019).

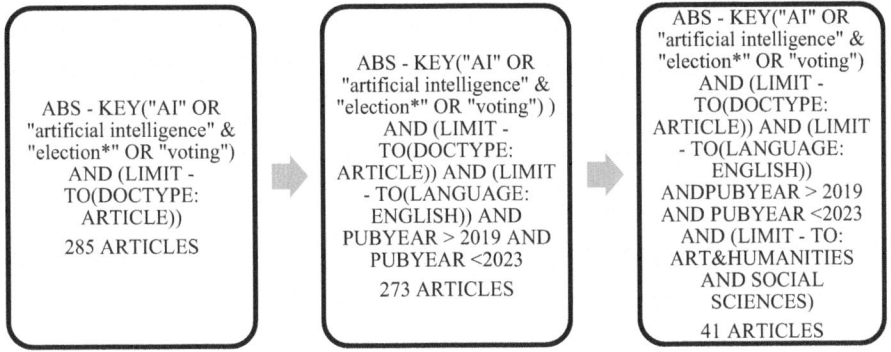

Fig. 2. Criteria of the article selection. Database: Scopus.

3 Analysis

3.1 Basic Analysis

In this section, the main information about the chosen papers will be presented. In the sample, there were 18 articles that were published in the following years: 2019–3 articles, 2020–2 articles, 2021–5 articles, 2022–5 articles, and 2023–3 articles. The subject of AI and elections in the context of fake news was present in 6 articles (Rocha Jr. et al. 2019; Biju and Gayathri 2023; Yankoski et al. 2021; Jungherr 2023; Chen et al., 2022; Aulia Rahman et al., 2022). This combination was justified by the recent phenomenon, where elections were influenced by fake news spread on social media. Examining the characterizations of AI put forth by the authors of the articles (Table 1) is crucial in order to correctly understand the research methodologies outlined in the reviewed articles. It is not surprising that some authors decided not to define AI, as they had difficulties providing a complex definition. Most papers provided a broad definition, encompassing a wide range of views on artificial intelligence. These approaches usually highlighted the ability to independently produce content without human interaction.

Table 1. Definition of artificial intelligence provided in researched articles.

Definition of AI	List of articles
machine learning models	(Kaspersen et al. 2022)
one of tools producing online fake news	(Biju and Gayathri 2023)
non-biological intelligent technologies	(Savaget et al. 2019)
Artificial Neural Networks	(Rosen 2023)
algorithms for automation	(Burgess 2022)
narrow AI trained on domain-specific data to perform domain-specific tasks	(Jungherr 2023)

3.2 Analysis of Research Methods

This section contains an overview of the research techniques and tools used in the sample. In Table 2, various methodological approaches are presented. It is important to distinguish Jungherr's framework. He described four areas in which AI may influence democracy. These were concentrated on different social group scales: individual level, group level, institutional level, and system level (Jungherr 2023). Such distinction provided the author with an opportunity to profoundly analyze the possible impact on democracy from various perspectives.

Table 2. Research methods applied in researched articles.

Research method	List of articles
Doctrinal legal study with a conceptual and comparative approaches	(Aulia Rahman et al. 2022)
Market analysis of the online sales areas for social bots and data crawling	(Assenmacher et al., 2020)
Literature review & case study	(Sidhu & Singhh 2021, Kane 2019; Savaget et al. 2019; Yankoski et al. 2021; Mainz et al. 2022; Ray 2021; García-Orosa 2021)
Literature review & creating blockchain system for fake-news prevention	(Chen et al., 2022)
Empirical study – quantitative approach based on data collected from a designed tool	(Kaspersen et al. 2022)
Literature review & conceptual framework	(Jungherr 2023; Burgess 2022)
Mixed methodology	(Savaget et al. 2019)
NLP / creating new framework based on NLP	(Yankoski et al. 2020; Biju and Gayathri 2023; Rosen 2023; García-Orosa 2021)

There was also mixed methodology provided in a paper focused on empowering political participation through AI (Savaget et al. 2019). The authors used bibliometric analysis, a systematic review of a sample of 721 publications, and a case study including interviews and participant observation, later analyzed through textual coding. All those methods led the authors to the creation of a framework of six main areas where AI technologies can empower civil society.

One paper included an empirical study involving 61 students and providing them with a specially developed tool to explore the impact of AI on voting.

(Kaspersen et al. 2022). The sample of the research is unique, as its aim is to study the opportunities and challenges for a CE approach in engaging students to understand and reflect on machine learning. It is not directly connected with the subject of elections. However, the unique tool provided for the research, called VotestratesML, is an application that enables students to build models that predict voters' behavior. VotestratesML is based on voter profile data from a survey of the Danish national election in 2015 (Kaspersen et al. 2022). In this case, researchers also used other methods: a literature review, a case study, a Design Research methodology, and an analysis of two methods of educating about machine learning models that predict election results.

It can be concluded that simple models based on machine learning already exist, so it is only a matter of time before we construct methods that will provide more accurate data about possible voting results.

Another application that was used for research on the topic of artificial intelligence and elections is Veirific.AI (Rocha Jr. et al. 2019). It is a tool used for checking news links, available on the Android system. The application was based on data mining to define the criteria for credible and fake news. Veirific.AI was tested during the Brazilian elections in 2018. During the 12-day test, the application received 373 links for verification. This study illustrates that the spread of fake news can be weakened by automated verification tools.

Two articles were based on data scraping (Yankoski et al. 2020; Biju and Gayathri 2023) and NLP methods combined with creating a new framework (Rosen 2023; García-Orosa 2021). Data scraping was conducted on social media in order to provide data for AI systems detecting disinformation (Yankoski et al. 2020) and for researching fake news spread by AI and bots in conflict zones in India (Biju and Gayathri 2023). This research method achieves satisfying results and should be further developed.

3.3 Influence on Elections

In the analyzed papers, most of the researchers agree that in the era of the AI rush, disinformation spread online can lead to social conflict (Biju and Gayathri 2023), possibly influence democratic elections (Yankoski et al. 2020) or weaken public institutions (Rocha Jr. et al. 2019). Going further, tools such as Facebook for politics, if empowered with AI solutions, can easily manipulate the political communication reaching the voters (Kane 2019). In order to prevent the negative impact of online disinformation, researchers work on verification tools that can easily check the credibility of each piece of information, including memes and videos (Yankoski 2020; 2021). However, it is always a two-sided battle in which manipulation (deep-fakes) and verification technologies are constantly evolving. In this context, in the future, AI literacy will be a crucial ability to fully participate in society (Kaspersen et al. 2022).

Additionally, AI can have desirable and undesirable effects on democracy. In the Pandora scenario (Savaget et al. 2019), AI can weaken democracy due to the possibility of centralizing and controlling all information, fake news, filter bubbles, etc. On the other hand, in the Jeffersonian scenario, AI can empower democracy by including marginalized groups, informing them about the political process, and giving them the opportunity to verify the news (Savaget et al. 2019). Moreover, artificial intelligence might open new possibilities for empowering civil society by researching data connected with political engagement, the will to vote, public expenses, etc. (Savaget et al. 2019).

4 Conclusions

Through this research, the author provided a holistic view of research methods that can be implemented for further examination of the relationship between AI and elections. Considering the dynamic evolution of AI-based solutions, there is a significant need to research the subject of AI's influence on elections and consider its implications for future legislation. Such research should be based on the artificial intelligence definition that will be stated by the author as a result of a definition review. There is a research gap in conducting empirical studies on the impact of AI on elections. The most profound understanding of the impact of artificial intelligence on elections is provided by the implementation of mixed methods, including bibliometric analysis, systematic review, case study, interviews, and participant observation. The presented methodological approach is limited because the literature review was confined to its keywords in the abstracts of articles. This could exclude articles with relevant content in the body but not mentioned in the abstract. Future research projects should concentrate on the possible consequences of implementing AI in technology that can lead to democracy and election crises, as there are already many articles covering theoretical frameworks in this area (Burgess 2022; Mainz et al. 2022). Artificial intelligence can be used to predict election results and provide recommendations for the politicians. Micro-targeted campaigns in social media, creating fake news and deepfakes are one of the most difficult challenges of the present and the future.

References

Assenmacher, D., Clever, L., Frischlich, L., Quandt, T., Trautmann, H., Grimme, C.:. Demystifying social bots: on the intelligence of automated social media actors. Soc. Media + Soc. **6**(3), 205630512093926 (2020). https://doi.org/10.1177/2056305120939264

Aulia Rahman, R., Nathanael Prabowo, V., David, A. J., Hajdú, J.: Constructing responsible artificial intelligence principles as norms: efforts to strengthen democratic norms in Indonesia and European Union. Padjadjaran J. Ilmu Hukum (Journal of Law) **9**(2), 231–252 (2022). https://doi.org/10.22304/pjih.v9n2.a5

Bareis, J., Katzenbach, C.: Talking AI into being: the narratives and imaginaries of national AI strategies and their performative politics. Sci. Technol. Hum. Values **47**(5), 855–881 (2022). https://doi.org/10.1177/01622439211030007

Biju, P.R., Gayathri, O.: Self-Breeding Fake News (2023)

Burgess, P.: Algorithmic augmentation of democracy: considering whether technology can enhance the concepts of democracy and the rule of law through four hypotheticals. AI & Soc. **37**(1), 97–112 (2022). https://doi.org/10.1007/s00146-021-01170-8

Chen, C.-C., Du, Y., Peter, R., Golab, W.: An Implementation of fake news prevention by blockchain and entropy-based incentive mechanism. Soc. Netw. Anal. Min. **12**(1), 114 (2022). https://doi.org/10.1007/s13278-022-00941-5

Farkas, J., Schou, J.: Post-truth discourses and their limits: a democratic crisis? In: Disinformation and Digital Media as a Challenge for Democracy (2020)

Field, A.: Mining the ambient commons: building interdisciplinary connections between environmental knowledge, AI and creative practice research. Interdisc. Sci. Rev. **47**(2), 185–198 (2022). https://doi.org/10.1080/03080188.2022.2036408

García-Orosa, B.: Disinformation, social media, bots, and astroturfing: the fourth wave of digital democracy. El Profesional de La Información, e300603 (2021). https://doi.org/10.3145/epi.2021.nov.03

Helm, J.M., et al.: Machine learning and artificial intelligence: definitions, applications, and future directions. Curr. Rev. Musculoskelet. Med. **13**(1), 69–76 (2020). https://doi.org/10.1007/s12178-020-09600-8

Jungherr, A.: Artificial intelligence and democracy: a conceptual framework. Soc. Media Soc. **9**(3), 20563051231186350 (2023). https://doi.org/10.1177/20563051231186353

Kane, T.B.: Artificial intelligence in politics: establishing ethics. IEEE Technol. Soc. Mag. **38**(1), 72–80 (2019). https://doi.org/10.1109/MTS.2019.2894474

Kaspersen, M.H., Bilstrup, K.-E.K., Van Mechelen, M., Hjort, A., Bouvin, N.O., Petersen, M.G.: High school students exploring machine learning and its societal implications: opportunities and challenges. Int. J. Child-Comput. Interact. **34**, 100539 (2022). https://doi.org/10.1016/j.ijcci.2022.100539

Kok, J.N.: Artificial intelligence: definition, trends, techniques and cases. Artif. Intell. **1**(270–299) (2009)

Krafft, P.M., Young, M., Katell, M., Huang, K., Bugingo, G.: Defining AI in policy versus practice. In: Proceedings of the AAAI/ACM Conference on AI, Ethics, and Society, pp. 72–78. ACM, New York (2020)

Mayring, P.: Qualitative content analysis: theoretical foundation, basic procedures and software solution (2014)

Moffitt, U., Rogers, L.O., Mzizi, Y., Charlson, E.: Race talk during the 2020 U.S. presidential election: emerging adults' critical consciousness and racial identity in context. J. Adolesc. Res. **39**, 1048–1085 (2022). https://doi.org/10.1177/07435584221145009

Mongrain, P.: Did you see it coming? Explaining the accuracy of voter expectations for district and (sub)national election outcomes in multi-party systems. Elect. Stud. **71**, 102317 (2021). https://doi.org/10.1016/j.electstud.2021.102317

Mainz, J.T., Sønderholm, J., Uhrenfeldt, R.: Artificial intelligence and the secret ballot. AI & Soc. (2022). https://doi.org/10.1007/s00146-022-01551-7

OECD: Recommendation of the Council on Artificial Intelligence (2019)

P, D., Simoes, S., MacCarthaigh, M.: AI and Core Electoral Processes: Mapping the Horizons (arXiv:2302.03774). arXiv. http://arxiv.org/abs/2302.03774 (2023)

Page, M.J., et al.: The PRISMA 2020 statement: an updated guideline for reporting systematic reviews. BMJ **372**, n71 (2021). https://doi.org/10.1136/bmj.n71

Ray, A.: Disinformation, deepfakes and democracies: the need for legislative reform. Univ. New South Wales Law J. **44**(3), 983–1013 (2021). https://doi.org/10.53637/DELS2700

Reis, J., Santo, P. E., Melao, N.: Impacts of artificial intelligence on public administration: a systematic literature review. In: 2019 14th Iberian Conference on Information Systems and Technologies (CISTI), pp. 1–7 (2019). https://doi.org/10.23919/CISTI.2019.8760893

Rocha Jr, D.B., Lins, A.J.D.C.C., De Souza, A.C. F., Libório, L.F.D.O., Leitão, A.H.D.B., Santos, F.H.S.: VERIFIC.AI application: automated fact-checking in Brazilian 2018 general elections. Braz. Journalism Res. **15**(3), 514–539 (2019). https://doi.org/10.25200/BJR.v15n3.2019.1178

Rethlefsen, M.L., et al.: PRISMA-S: an extension to the PRISMA statement for reporting literature searches in systematic reviews. Syst. Rev. **10**(1), 39 (2021). https://doi.org/10.1186/s13643-020-01542-z

Rosen, Z.P.: A BERT's eye view: a big data framework for assessing language convergence and accommodation. J. Lang. Soc. Psychol. **42**(1), 60–81 (2023). https://doi.org/10.1177/0261927X221095865

Savaget, P., Chiarini, T., Evans, S.: Empowering political participation through artificial intelligence. Sci. Pub. Policy **46**(3), 369–380 (2019). https://doi.org/10.1093/scipol/scy064

Schuett, J.: Defining the scope of AI regulations (2022)

Secinaro, S., Calandra, D., Secinaro, A., Muthurangu, V., Biancone, P.: The role of artificial intelligence in healthcare: a structured literature review. BMC Med. Inform. Decis. Mak. **21**(1), 125 (2021). https://doi.org/10.1186/s12911-021-01488-9

Sharma, G.D., Yadav, A., Chopra, R.: Artificial intelligence and effective governance: a review, critique and research agenda. Sustain. Fut. **2**, 100004 (2020). https://doi.org/10.1016/j.sftr.2019.100004

Sidhu, B. K., Singhh, A.: Cybersecurity Regulatory Landscape in India: Digitisation on the Hook? **56**(38) (2021)

Stark, L., Crawford, K.: The work of art in the age of artificial intelligence: what artists can teach us about the ethics of data practice. SS **17**(3/4), 442–455 (2019). https://doi.org/10.24908/ss.v17i3/4.10821

Verma, S., Sharma, R., Deb, S., Maitra, D.: Artificial intelligence in marketing: systematic review and future research direction. Int. J. Inf. Manage. Data Insights **1**(1), 100002 (2021). https://doi.org/10.1016/j.jjimei.2020.100002

Yankoski, M., Scheirer, W., Weninger, T.: Meme warfare: AI countermeasures to disinformation should focus on popular, not perfect, fakes. Bull. Atomic Sci. **77**(3), 119–123 (2021). https://doi.org/10.1080/00963402.2021.1912093

Yankoski, M., Weninger, T., Scheirer, W.: An AI early warning system to monitor online disinformation, stop violence, and protect elections. Bull. Atomic Sci. **76**(2), 85–90 (2020). https://doi.org/10.1080/00963402.2020.1728976

AI in Society and Legal Aspects

On the Legal Aspects of Responsible AI: Adaptive Change, Human Oversight, and Societal Outcomes

Daria Onitiu[1(✉)], Vahid Yazdanpanah[2], Adriane Chapman[2], Enrico Gerding[2], Stuart E. Mid-dleton[2], and Jennifer Williams[2]

[1] Oxford Internet Institute, 1 St Giles', Oxford OX1 3JS, UK
daria.onitiu@oii.ox.ac.uk
[2] University of Southampton, University Road, Southampton SO17 1BJ, UK

Abstract. This paper discusses the ways in which complexity and degrees of autonomy in AI-based medical devices (AIaMD) may challenge the safety and performance of software for EU regulatory alignment and responsible AI regarding AI-induced harms. It examines the EU Commission proposals for an AI Liability Directive and a revised Product Liability Directive to identify two research challenges. These challenges relate to identifications of "defects" arising from algorithmic change and degrees of human oversight. Some suggestions will be made in how they can be addressed through causal modelling, counterfactuals, and responsibility reasoning.

Keywords: EU liability · product liability · AI- based medical devices · AI

1 Introduction

Developments in Artificial Intelligence (AI) techniques in healthcare – from advanced computer vision methods and deep learning in medical imaging to machine learning techniques for personalised medicine – raise important questions regarding who and to what extent could be liable for the consequences of their harms during product life cycle[1, pp. 14–15], [2–4]. Of particular importance are some special characteristics of AI- based medical devices (AIaMD); from concerns surrounding bias, opacity, and human oversight, to some models being able to "continuously learn" on real-world data [5, p. 11]. In this article, we briefly introduce how this specific complexity in AIaMD produces some regulatory tension with the strict liability and fault- based liability regime currently proposed in the revised Product Liability Directive (PLD) and the AI Liability Directive (ALD) [6, 7]. We discuss two research challenges illuminating on a tension of implementing safety and performance requirements within proposals for a revised liability framework in the EU and pertaining to AI-based innovations. Our recommendations concentrate on technical safeguards for traceability and modelling human-agent reasoning intending. These recommendations intend to supplement future policy efforts as well as direct on identifying "design defects" and "non-compliance" with some duties in the "high-risk" obligations in the upcoming Regulation on Artificial Intelligence (AI Act) [8].

The direction of EU governance for AI is currently at a crucial point; with the EU Commission's proposal for an AI Act progressing and entering trilogue negotiations [8], the ambitious efforts within the EU to engage into technical standard setting for "high-risk systems" [9], the revision of the product liability framework, as well as the introduction of a new framework on procedural harmonisation of fault-based torts [6, 7].

The proposals for a revised PLD and the ALD intend to illustrate a steppingstone for a more comprehensive AI regulation focusing on the consequences of AI- induced harms. Whilst both frameworks are still at the early stages of development [10], their inherent link to the AI Act proposal, as well as fundamental concepts on the role of software-related harms are an important source for us to issue further comment. In doing so, we intend to focus on two specific aspects within this legislative net on EU liability; one is the notion of "defect" in the PLD, and second, is the "presumption of causality" focusing on the provisions of human oversight in the AILD and referring to the AI Act [8].

In this work, we refer to AI and autonomous systems to describe liability issues for special to continuously updated systems. Indeed, we recognise different degrees of "change" in that a medical AI system may either be "locked", allow for "incremental learning" or are "adaptive" [11, p. 12], [12]. Focusing on systems that are "adaptive", what is crucial here is that these decisions are made with little to no direct human intervention [13, p. 6]. Following this narrative, we identify two research challenges which are relevant to the way future regulatory responses will ensure alignment between EU (sectoral) legislation, the AI Act and responsible AI. These are (i) the way a manufacturer may intervene with adaptive change when AIaMDs exhibit a degree of autonomy for the identification of "defects" in the PLD; and (ii) the role of human-AI interaction in usability and risk management when distributing fault-based claims using the ALD's "presumption of causality" for human oversight.

Moving forward, we make two recommendations in how causal modelling, counter-factuals, and responsibility reasoning, can supplement performance and safety assurance of AIaMD with a view to addressing these research challenges. The paper recommends human-agent reasoning approaches, which are just a subset of approaches for AI assurance that are being actively researched and explored today [14]. Assurance methods can be applied to any part of the AI lifecycle, including data management, model learning, model verification and model deployment. For model verification prior to deployment approaches include probabilistic verification methods, generative adversarial networks, and combinatorial testing. For model assurance of post-deployment behavior, monitoring approaches are very important, specifically focusing on model inputs, outputs, and the environment in which the model operates. Techniques include built-in tests, fallback model safe states and well-defined tolerance thresholds within which model outputs must stay.

2 Why Contours of EU Liability do not Address Challenges Pertaining to AIaMD

Imagine an AIaMD which employs deep learning algorithms for the interpretation of Chest X-rays and intended to be used for assistive diagnosis of pneumonia (see also, [15]). Developers validating the device based on a significant number of labelled medical images for training and evaluation of the algorithm were able to show the system's impressive accuracy to distinguish between features in an image, which would outperform the judgement of a radiologist [16]. Nevertheless, this system when deployed into a healthcare environment could have serious implications for patient safety: from possible risks based on the opacity of the model leading up to errors in use, degrading in performance once exposed to real-world data, to learning from spurious correlations in the training data ([17, p. 1202], [18–20]). Especially relevant to AIaMD is that such risks could result in adverse outcomes that could produce patient harm, including misdiagnosis and consequently mistreatment of the patient.

These problems on the EU regulation of AIaMD for patient safety, touching upon the degrees of safety and performance assurances within the Medical Device Regulation [21, p. 3], incorporate important liability questions. Questions arise to what extent developers can be liable for any possible "defects" arising from "differential performance"; i.e., deviations of the algorithm from (pre-defined) parameters. In other words, to what extent does the manufacturer including developers maintain control for this specific type of adaptive changes causing harm? (see also, [22]).

In addition, if we assume that the AIaMD assists healthcare professionals regarding the diagnosis and treatment of different stages of pneumonia, then questions for minimising use errors and ensuring risk management on the one hand and providing for human oversight on the other hand, are important elements producing liability gaps. For example, if a human agent overrides the output of a fully and opaque autonomous system, does that provide for the liability of the user of the "high-risk system"? (see also, [11, p. 2]).

The latest proposals for two EU Liability Directives – the PLD and ALD– seek to resolve some tensions arising from the challenge to allocate liability regarding software-related failures, as well as tensions arising from burden of proof for fault-based claims. The revised PLD includes in its material scope medical device software and AI systems ([23, Art. Recital 12], see also, [24, p. 826]). Another objective of the proposed EU Liability framework is to incorporate the special characteristics of AI- algorithmic complexity, autonomy, and opacity- which would make it substantially difficult for victims to succeed in a liability claim based on a "lack of compliance under Union or national law" [7, Art. Explanatory Memorandum]. In this respect, The ALD, extending on the provisions in about "high-risk" systems in the AI Act, allows the victim to claim damages regarding harm caused by medical AI systems to safety and/or fundamental rights (compare with [3]).

The extent to which these two legislative frameworks successfully address gaps on liability still leaves some questions open. Focusing on our example of AIaMD used for the detection of pneumonia, it could be argued that a "defect" assumes that there is a well-defined standard from which errors in design and performance can be judged [25, p. 24]. But such well-defined standards are not currently under the remit of AI regulation, and instead are currently defined based on academic research or in-house by manufacturers within their own product safety departments, if at all [26]. Similarly, if a user (over)-relies on the AI system's opaque output then it is unclear to what extent the ALD would provide any victims with redress under this framework, including the ability to know if there have been any victims [3], [11, p. 2].

Our analysis proceeds with the identification of two research challenges on the safety and performance of AIaMD stimulating future discussion on a comprehensive the EU Liability regime. First, we argue that making the developer strictly liable for software-related failures and hazardous situations arising from the use of AIaMD after deployment, needs to be accompanied with further specifications for developers to effectively intervene with adaptive learning. Second, we argue that the ALD currently provides the wrong incentives to "prove fault" only on the basis of incorrect specifications for human oversight [7, p. 4 (2); 4 (3)], leaving out an insight into the elements for "providers" to justify the AIaMD's risk profile and "users" to follow the necessary competence for intervention.

2.1 Research Challenge 1: Lack of Specifications for Manufacturers Justifying Interventions with the System's Adaptive Changes

Our first statement of why the proposed EU Liability framework does not fully capture risks surrounding design and deployment of AIaMD is based on the lack of guidance for directing contours of continuous and adaptive change when the system exhibits a degree of autonomy. As a starting point, enabling change through updates is an important aspect of "software maintenance" being a necessary component to ensure patient safety [27, Sec. 6]. In this regard, many currently approved AIaMDs employ models that are "locked", providing the same "output on a given set of inputs" [28, p. 3], [29, p. 3]. "Batch-learning" refers to instances where the algorithm does retrain itself incrementally after seeing a new batch of training data [30, p. 30]. In these given instances we can argue that the manufacturer retains some form of control regarding the extent software updates are pushed through the lifecycle. This reasoning of control extends to "defects" arising from the lack or improper controls including software updates altering performance, safety, and functionality after deployment during "its lifespan" [6, p. 10 (2) (b)-(c)], [31, Art. Amendments 103–104].

However, there is a lot of potential in algorithmic models that are programmed to internally modify their algorithms for a new output based on real-world data [13, p. 8]. These models are adaptive in that these algorithms continuously learn and change their performance [17, p. 1202]. Whilst these models could be useful to provide more "timely.

recommendations" based on real-world experience [32, p. 678], these characteristics could pose additional issues for validation and oversight [13, p. 71].

One issue is that the adaptive nature of algorithms poses a challenge for manufacturers to recognise "significant changes" and/or performance degradation impacting patient safety [33, pp. 32–33]. This in turn can have significant implications for the extent to which "unpredictable" changes would be classified as a "defect" under the revised PLD leading to "material damage" [11, p. 4], [34, p. 28]. This notion of "defectiveness" flows from the failure to disclose relevant evidence about the product, lack of compliance with mandatory safety legislation, such as the EU Medical Device Regulation, or by virtue of an "obvious malfunction" [6, Art. 9 (2) (a)-(c); 8 (1); 9 (3)]. As highlighted by Borges [4, pp. 4–5], once the manufacturer's control effectively relinquishes regarding the operation of the system, then "defectiveness" has to be inferred from the system's behaviour. To establish liability then would "require defining property as the ability to behave in a certain way in a certain situation or not to show a certain behaviour" [35, p. 35]. In this respect, Article 6 (1) (c) of the revised PLD, when read in conjunction with its version from June 2023, establishes that a product could be "considered defective when it does not provide the safety which the public at large is entitled to expect, taking all circumstances into account, including…the effect on the product of any ability to continue to learn" "after it is placed on the market or put into service" ([6, Art. 6 (1) (c)], [23, Art. 6 (1) (c)]; compare with, [31, Art. Amendment 70]. The Parliamentary report further refines this definition by stating that the "effect of the product" includes "any ability to acquire new features or knowledge after it is placed on the market or put into service" [8, Art. Article 6 (1) (c); Amendment 74].

A specific hurdle for regulators to effectively adapt this notion of "defectiveness" for AIaMD concerns those exact specifications tracing adaptive change. The evaluation of AIaMD needs to be subject to pre-defined parameters whilst the variability of risk occurs *within* those dynamic changes. This is effectively a problem that lies at the heart of the verification and validation of AIaMD; once the manufacturer evaluates the medical device's specific and intended use, the next step is monitoring the extent that the system is operating within "a set of underlying assumptions" [36, p. 442].

Hence, an important challenge in a comprehensive legal responsibility framework of AI is determining the manufacturer's formalisation of the system's safety and performance when interacting with various stakeholders on the ground, such as users, healthcare professionals, and patients. Indeed, regulatory developments are currently dealing with the problems associated with the regulation of adaptive algorithms in medicine. This can be seen in guidance by the U.S Food & Drug Administration [29], current efforts by the UK Medicines & Healthcare products Regulatory Agency (MHRA) [37], as well as a draft reflection piece by the European Medicines Agency (EMA) [38]. In this regard, the EMA thought-piece enumerates "thresholds for model performance" that are needed for manufacturers to monitor degradation and failure modes of algorithms [38, Sec. 2.4.6].

Moreover, "the presidency draft compromise proposal" and "the Council's negotiating position" clarify important notions on "defectiveness" which may entail "upgrades and updates of machine learning algorithms" ([23, Art. Recital 37], [39, Art. Recital 37]; see also, [8, Art. Amendment 39]. What is missing; however, is placing these considerations into a wider context on how manufacturers justified any residual risks before and after deployment.

This is because the notion of "defect" in the revised PLD assumes the manufacturer's level of control regarding anticipated modifications, considering Article 4 (10a) (a) in the revised PLD and thresholds for modification under the Medical Device Regulation [23, Art. 4 (10a) (a)], [40, Art. 7 (d)]. The question, therefore, is which level of modifications justify deviations within a system's acceptable risk policy after deployment and recognising an extent of diminishing control. What follows is that, in addition to the extent that a system re-trains dynamically, we need further specifications in *how* manufacturers can justify an intervention with underlying assumptions surrounding adaptive change. That is, most specifications that are "pre-determined" by the manufacturer from the outset are likely to be either inconsistent or incomplete with adaptive algorithms [17, p. 1203]. In addition, if we assume that the manufacturer producing a risk management plan considering performance thresholds for adaptive systems, this has to be based on clear criteria for manufacturers to intervene with a pre-determined policy. Therefore, what matters is for manufacturers to be able to justify any residual risks during the product life cycle and articulate how risk control and mitigation maintain the manufacturer's control.

2.1.1 The Role of Causal Reasoning for Adaptive Change

A promising way forward for addressing challenges around intervening with change for continuously learning AIaMD is to leverage the potentials of computational causal models and reasoning. Causal modelling and reasoning methods, such as those developed by Judea Pearl [41] and Joseph Halpern [42], can provide useful frameworks for modelling and anticipating the impacts of algorithmic changes to AIaMD. These methods allow developers to map out the potential chains of cause-effect relationships stemming from modifications to the AI system's algorithm. By forecasting how various causal factors may lead to harmful outcomes, developers can take steps to avoid or mitigate these risks proactively. Causal models that accurately represent the dynamics of the AI system, its real-world deployment context, and interactions with human users can enable explanatory insights about how and why unintended consequences may arise. Specifically, formal causal reasoning can supplement standard verification and validation protocols by identifying probable failure points or risks. In this way, causal modelling supports the design of safer AI systems and responsibility frameworks that account for complex sociotechnical interactions. Integrating these techniques can strengthen technical specifications for dynamic changes and manufacturer oversight of medical AI.

Furthermore, causal models can assist stakeholders in determining when and how to appropriately intervene whilst tracking risks of performance related harms. By modelling the dynamics of the AI system, developers can use causal models to identify specific adaptive changes and significant changes in continuously learning algorithms. This enables them to focus interventions on the most impactful areas, whether through altering technical specifications, adjusting training data, or implementing human oversight mechanisms. Causal reasoning allows for targeted interventions that avoid unnecessary restrictions on beneficial adaptations of the AI, while still providing safeguards against harmful impacts on health, safety. In this way, causal models support nuanced and selective interventions on high-risk changes, guided by causal understanding of how different modifications contribute to various intended and unintended consequences.

2.2 Research Challenge 2: Identifying the Parameters for Risk Management and Usability of AIaMD

The second research challenge, focusing on the interpretation of the ALD proposal, is aligned with the AI Act's parameters for human oversight regarding AI used for decision-support. An important aspect of the proposal for an ALD is that it establishes a framework that "harmonise[s] non-contractual civil liability rules" ([43, p. 5]; see also, [44]). In doing so, it intends to ensure that the victim, such as the individual patient harmed by the output of the AIaMD detecting pneumonia, to have recourse to the developer, the provider or user's and prove "fault" during the use of the "high-risk" system that may have caused damage [7, Art. 4], [21, p. 5].

In this regard, Article 4 (1) entails a rebuttable "presumption of a causal link in the case of fault" [7, Art. 4 (1)]. For example, if the provider (i.e., manufacturer or a person placing the product on the market) or "user" (i.e., a person using the AI system under the provider's authority) do not comply with the relevant provision of "human oversight" in the AI Act, then Article 4 implies that fault can be inferred from the fault on the provider or user based on non-compliance of provision relating to "high-risk" systems under the AI Act and based on the "the output produced by the AI system or the failure of the AI system to produce an output"[7, Art. 4(1)(b); 4(2); 4(3); Recital 25]. Returning to the earlier example of AI-based pneumonia detection, this scenario would be exemplified by a qualified medical professional who received the relevant training for instructions of use whilst using the tool in a manner that later causes the algorithm to adaptively learn from incorrect or incomplete inputs during real-time learning updates. The question is how to distribute the levels of human oversight and concerns of risks of (over)reliance during the system's lifecycle (see also, [21, pp. 7–8]).

The defendant may rebut the presumption in Article 4(1) "by showing that its fault could not have caused the damage" [7, Art. Recital 20; 3(5); 4(1); 4(7)]. In addition, Article 4(4) argues that the presumption does not apply in instances where the defendant successfully argues that "sufficient evidence and expertise is accessible for the claimant to prove the causal link" [7, Art. 4(4)]. Furthermore, the fact that the presumption is rebuttable highlights that it must have been "reasonably likely" on a case-by-case basis that the fault occurred based on non-compliance with a duty that had an impact on the system's output [7, Art. 4(1) (b)].

Non-compliance with the duty of human oversight, operationalised in a safety critical domain, opens an interesting discussion on how liability is distributed between various stakeholders. This is because human oversight is an aspect of usability for risk management as well as following instructions of use (see, [45]). To effectively implement this requirement for AIaMD requires more specifications of the limits of the system's intended use [46, p. 171]. For instance, a consultation by the BSI and AAMI further highlights that device complexity and autonomy require human-in-the-loop testing as well as evaluations in how user interface design can introduce "automation bias" [45, Sec. 5.9.1–5.9.5]. Having said that; however, the lack of specifications for human oversight of AIaMD on the EU regulatory level currently leaves manufacturers considerable leeway to justify the limitations of a "high-risk" system interacting with stakeholders on the ground [46, p. 180].

This also opens up gaps in of how risk management and usability informs claims of suspected harm against the provider and user. Article 4 (2) of ALD whilst highlighting the "results of the risk management system" as a factor for constructing the "rebuttable presumption of causality" poses issues for interpreting usability at a level of complexity of the intended environment and user of AIaMD [7, Art. Article 4 (2)]. This reasoning is based on its construction of "fault" as a "failure of the AI system to produce an output" which focus on deviations of specifications based on inadequate risk controls that lead to an erroneous assessment of the "overall residual risk" [45, Sec. 8.1]. A clear example of an incorrect specification is for developers to not include "warnings" or "specific training" for users when these elements would be a necessary component for using a AIaMD for assistive diagnosis and maintaining the benefit-risk profile [47, Sec. 7].

In other instances, however, it is less clear how the effects of specifications, influencing over-reliance and possibly producing "automation bias" [48], would trigger the "presumption of causality" for aspects relating to EU product safety assurances relating to AIaMD. This would require the identification of inconsistencies on the evaluation of usability and risk management, such as for the victim showing an imbalance in the overall benefit-risk ratio. A claim against the provider of the AIaMD for non-compliance with human oversight is currently narrowed to the developer's specifications of risk controls for usability. Article 4 (2) of the ALD proposal only allows for proving fault on the basis of incorrect specifications; not how limitations of specifications of usability caused *wrong incentives*, when users are interacting with the system [7, Art. 4 (2)]. This is because Article 4 (2) elaborates on the connection of fault without necessarily tapping into the manufacturer's justification of the overall residual risk [7, Art. 4 (2)].

Turning to the limitations of Article 4 (3) of the ALD proposal, Hacker makes an important point that interrogates the strong link "between the fault and the output… [not the extent of human oversight] after the AI output is produced" [25, p. 36]. This is further seen in Recital 25 of the ALD illuminates how "fault" is usually tied to the boundary specifications such as the breach of "the perimeter of operation of the AI system" [7, Art. Recital 25]. Additionally, Recital 15 highlights that claims against the user for human omissions implementing the AI output means that responsibility for the damage needs to be traced back to "an output or the failure to produce an output by an AI system through the fault of a person" [7, Art. Recital 15].

Without guidance on the appropriate level of interaction between the human and AI, a victim will have great difficulty to show the consistency between usability specifications on the intended user and the oversight shaped by the AI output. In particular, it opens up a tension for the interpretation of Articles 4(2) and 4(3) of the ALD proposal that is limited to the user simply "implementing the AI system's output" and the risks of overreliance detached from the safety and effectiveness of a "high-risk system" [7, Art. 4(2)–4(3)].

2.2.1 Counterfactual Scenarios Illuminating on Degrees of Human-AI Interaction

Modelling counterfactual scenarios can support stakeholders in reasoning about the avoidance potential of different agents and whether they could have prevented outcomes producing a clear link between fault and non-compliance that caused harm [49, 50]. By examining alternate possibilities along the timeline, not just future risks, these models elucidate the range of actions manufacturers, providers, and users could have taken at each decision point. Or it could simulate how increased human monitoring and overrides at the deployment stage could have prevented improper implementation of the system's output. Evaluating these counterfactuals helps determine if and where human oversight failed or could be enhanced. This analysis enables clearer delineation of responsibility by revealing who had the knowledge and capacity to avoid harms at various stages [51].

Building on counterfactual models, computational techniques for responsibility assignment also hold promise [52–54]. These approaches formally analyse the responsibilities of human and AI agents given counterfactual trajectories. The modelling considers agents' available actions, knowledge of likely outcomes. By simulating adherence to these norms, the model can assess and oversight. Encoding factors like safety rules, computational models can verify whether following the requirements could have changed outcomes. Verifying responsibility through this computational approach accounts for complex sociotechnical dynamics between manufacturers, health providers, and end users.

Importantly, formal computational models allow distinguishing graded levels of responsibility, from causal contribution to harm [55]. Whereas causal models capture roles in producing an outcome, computational responsibility modelling also verifies awareness of avoidability. For instance, these techniques could determine if a harmful AI outcome was due to reasonably foreseeable misuse. This enables moving to nuanced designations from noncompliance to full culpability. In this way, computational techniques could strengthen assessment of human oversight and clarify ambiguity around legal liability for harms involving AI systems.

3 Concluding Thoughts

This paper illustrates a snapshot on some problems shaping EU regulators' effort for responsible AI and governance. It provides an understanding of the regulation of AIaMD which is based on their special characteristics and its specific benefit-risk profile to produce AI-induced harms. Whilst our discussion of the proposed EU Liability framework is not comprehensive, it picks up two important research challenges that evolve around the ongoing challenges for regulatory reform on the safety and performance of AIaMD. These are (1) problems of monitoring dynamic and adaptive change including continuously learning algorithms and (2) issues to specify human oversight of AIaMD used for decision-support.

An important limitation of our work is that our comments are clearly focused on the EU regulatory landscape and currently ongoing efforts by EU regulators. In this regard, future standard-setting could evolve horizontally within the AI Act and following an approach that is "assurance-based", as well as entail sector-specific standards for AIaMD whilst considering the Medical Device Regulation [56, p. 43], [57, p. 65]. Our findings, whilst not providing an exhaustive picture to the design and use of AIaMD, indicates that these issues for responsible AI are cross-sectoral, being relevant for the interpretation of notions pertaining to EU liability and finally, regulatory alignment within a legislative climate that grapples with providing technical standard setting, legal certainty and keeping up with technological developments.

Acknowledgements. This work was supported by the UK Engineering and Physical Sciences Research Council (EPSRC) through the Trustworthy Autonomous Systems Hub (EP/V00784X/1), a Turing AI Acceleration Fellowship on Citizen-Centric AI Systems (EP/V022067/1) and the UKRI Research Node on Trustworthy Autonomous Systems Governance and Regulation (EP/V026607/1). In addition, this work was supported by the Natural Environment Research Council (NE/S015604/1) and Economic and Social Research Council (ES/V011278/1), as well as the National Institute for Health and Care Research Southampton Biomedical Research Center (ISBRC-1215–20004). Finally, this work was supported through research funding by the Wellcome Trust (223765/Z/21/Z), Solan Foundation (G-2021–16779) and the Department of Health and Social Care (via the AI Lab at NHSx) and Luminate Group within the Trustworthiness Auditing for AI project and the Governance of Emerging Technologies research programme (GET) at the Oxford Internet Institute, University of Oxford.

References

1. Regulatory Horizons Council (RHC), 'RHC report on the regulation of Artificial Intelligence as a Medical Device' (2022). https://www.gov.uk/government/publications/regulatory-horizons-council-the-regulation-of-artificial-intelligence-as-a-medical-device. Accessed 20 Apr 2023
2. Price II, W.N., Gerke, S., Cohen, I.G.: Liability for use of artificial intelligence in medicine. Rochester, NY, 20 May 2022. https://doi.org/10.2139/ssrn.4115538
3. Ziosi, M., Mökander, J., Novelli, C., Casolari, F., Taddeo, M., Floridi, L.: The EU AI liability directive: shifting the burden from proof to evidence. Rochester, NY 06 June 2023. https://papers.ssrn.com/abstract=4470725. Accessed 09 June 2023
4. Borges, G.: Liability for AI systems under current and future law: an overview of the key changes envisioned by the proposal of an EU-directive on liability for AI. Comput. Law Rev. Int. **24**(1), 1–8 (2023). https://doi.org/10.9785/cri-2023-240102
5. Artificial Intelligence Medical Device Working Group, 'Machine Learning-enabled Medical Devices: Key Terms and Definitions', International Medical Device Regulators Forum (IMDRF), IMDRF/AIMD WG/N67 (2022). https://www.imdrf.org/sites/default/files/2022-05/IMDRF%20AIMD%20WG%20Final%20Document%20N67.pdf
6. Proposal for a DIRECTIVE OF THE EUROPEAN PARLIAMENT AND OF THE COUNCIL on liability for defective products (2022). https://eur-lex.europa.eu/legal-content/EN/TXT/?uri=CELEX%3A52022PC0495. Accessed 31 July 2023
7. Proposal for a DIRECTIVE OF THE EUROPEAN PARLIAMENT AND OF THE COUNCIL on adapting non-contractual civil liability rules to artificial intelligence (AI Liability Directive) (2022). https://eur-lex.europa.eu/legal-content/EN/TXT/?uri=CELEX%3A52022PC0496. Accessed 30 July 2023
8. DRAFT Compromise Amendments on the Draft Report Proposal for a regulation of the European Parliament and of the Council on harmonised rules on Artificial Intelligence (Artificial Intelligence Act) and amending certain Union Legislative Acts (COM(2021)0206 –C9 0146/2021 –2021/0106(COD)) (2023). https://www.europarl.europa.eu/meetdocs/2014_2019/plmrep/COMMITTEES/CJ40/DV/2023/05-11/ConsolidatedCA_IMCOLIBE_AI_ACT_EN.pdf. Accessed 16 Oct 2023
9. EU Commission, Draft standardisation request to the European standardisation organisations in support of safe and trustworthy artificial intelligence. https://ec.europa.eu/docsroom/documents/52376. Accessed 28 July 2023

10. Bertuzzi, L.: EU Council clarifies liability rules for software updates, machine learning. www.euractiv.com. https://www.euractiv.com/section/digital/news/eu-council-clarifies-liability-rules-for-software-updates-machine-learning/. Accessed 03 Aug 2023
11. Duffourc, M.N., Gerke, S.: The proposed EU directives for AI liability leave worrying gaps likely to impact medical AI. npj Digit. Med. **6**(1), 1 (2023). https://doi.org/10.1038/s41746-023-00823-w
12. Gepperth, A., Hammer, B.: Incremental learning algorithms and applications. In: European Symposium on Artificial Neural Networks (ESANN), Bruges, Belgium (2016). https://hal.science/hal-01418129. Accessed 21 Aug 2023
13. BSI and AAMI, 'MACHINE LEARNING AI IN MEDICAL DEVICE: Adapting Regulatory Frameworks and Standards to Ensure Safty and Performance'. https://www.bsigroup.com/en-US/medical-devices/resources/Whitepapers-and-articles/machine-learning-ai-in-medical-devices/. Accessed 21 Apr 2023
14. Ashmore, R., Calinescu, R., Paterson, C.: Assuring the machine learning lifecycle: desiderata, methods, and challenges. ACM Comput. Surv. **54**(5), 111:1–111:39 (2021). https://doi.org/10.1145/3453444
15. Nadis, S.: Using AI and old reports to understand new medical images: scientists employ an underused resource — radiology reports that accompany medical images — to improve the interpretive abilities of machine learning algorithms. MIT CSAIL. https://news.mit.edu/2021/using-ai-and-old-reports-understand-new-medical-images-0927
16. Topol, E.J.: High-performance medicine: the convergence of human and artificial intelligence. Nat. Med. **25**(1), 1 (2019). https://doi.org/10.1038/s41591-018-0300-7
17. Babic, B., Gerke, S., Evgeniou, T., Cohen, I.G.: Algorithms on regulatory lockdown in medicine. Science **366**(6470), 1202–1204 (2019). https://doi.org/10.1126/science.aay9547
18. Andersen, T.O., Nunes, F., Wilcox, L., Coiera, E., Rogers, Y.: Introduction to the special issue on human-centred AI in healthcare: challenges appearing in the wild. ACM Trans. Comput.-Hum. Interact. **30**(2), 25:1–25:12 (2023). https://doi.org/10.1145/3589961
19. Heaven, W.D.: 'Google's medical AI was super accurate in a lab. Real Life was a Different Story. | MIT Technology Review', MIT Technology Review. https://www.technologyreview.com/2020/04/27/1000658/google-medical-ai-accurate-lab-real-life-clinic-covid-diabetes-retina-disease/. Accessed 04 Aug 2023
20. Gichoya, J.W., et al.: AI recognition of patient race in medical imaging: a modelling study. Lancet Digit. Health **4**(6), e406–e414 (2022). https://doi.org/10.1016/S2589-7500(22)00063-2
21. Buiten, M., de Streel, A., Peitz, M.: The law and economics of AI liability. Comput. Law Secur. Rev. **48**, 105794 (2023). https://doi.org/10.1016/j.clsr.2023.105794
22. Sullivan, H.R., Schweikart, S.J.: Are current tort liability doctrines adequate for addressing injury caused by AI? AMA J. Ethics **21**(2), 160–166 (2019). https://doi.org/10.1001/amajethics.2019.160
23. Proposal for a Directive of the European Parliament and of the Council on liability for defective products Mandate for negotiations with the European Parliament (2023). https://data.consilium.europa.eu/doc/document/ST-10694-2023-INIT/en/pdf
24. de Graaf, T., Veldt, G.: The AI act and its impact on product safety, contracts and liability. Eur. Rev. Priv. Law **30**(5) (2022). Accessed 01 Aug 2023
25. Hacker, P.: The European AI liability directives -- critique of a half-hearted approach and lessons for the future. arXiv 28 July 2023. https://doi.org/10.48550/arXiv.2211.13960
26. Williams, J., Pizzi, K., Das, S., Noe, P.-G.: New challenges for content privacy in speech and audio. In: 2nd Symposium on Security and Privacy in Speech Communication, pp. 1–6 (2022). https://doi.org/10.21437/SPSC.2022-1
27. BS EN 62304:2006+A1:2015 Medical device software. Software life-cycle processes

28. Gerke, S., Babic, B., Evgeniou, T., Cohen, I.G.: The need for a system view to regulate artificial intelligence/machine learning-based software as medical device. npj Digit. Med. **3**(1), 1 (2020). https://doi.org/10.1038/s41746-020-0262-2
29. U.S Food & Drug Administration, 'Proposed Regulatory Framework for Modifications to Artificial Intelligence/Machine Learning (AI/ML)-Based Software as a Medical Device (SaMD) - Discussion Paper and Request for Feedback', FDA (2019). https://www.fda.gov/medical-devices/software-medical-device-samd/artificial-intelligence-and-machine-learning-software-medical-device. Accessed 20 Apr 2023
30. Ordish, J., Murfet, H., Hall, A.: Algorithms as medical devices. https://www.phgfoundation.org/report/algorithms-as-medical-devices. Accessed 16 Apr 2023
31. REPORT on the proposal for a directive of the European Parliament and of the Council on liability for defective products (2023). https://www.europarl.europa.eu/doceo/document/A-9-2023-0291_EN.html
32. Petersen, C., et al.: Recommendations for the safe, effective use of adaptive CDS in the US healthcare system: an AMIA position paper. J. Am. Med. Inform. Assoc. **28**(4), 677–684 (2021). https://doi.org/10.1093/jamia/ocaa319
33. Lekadir, K., Quaglio, G., Tselioudis, A., Gallin, C.: Artificial intelligence in healthcare: applications, risks, and ethical and societal impacts | Digital Skills & Jobs Platform. https://digital-skills-jobs.europa.eu/en/inspiration/research/artificial-intelligence-healthcare-applications-risks-and-ethical-and-societal. Accessed 20 Apr 2023
34. Directorate-General for Justice and Consumers (European Commission), Liability for artificial intelligence and other emerging digital technologies. LU: Publications Office of the European Union (2019). https://doi.org/10.2838/573689. Accessed 01 Aug 2023
35. Borges, G.: AI systems and product liability. In: Proceedings of the Eighteenth International Conference on Artificial Intelligence and Law. ICAIL 2021, pp. 32–39. Association for Computing Machinery, New York, NY, USA (2021). https://doi.org/10.1145/3462757.3466099
36. Neelke, D.P.M.D. (ed.) Francien Dechesne and Tijn Borghuis on Verification and Validation in Engineering. In: The Routledge Handbook of the Philosophy of Engineering, Routledge, New York (2020). https://doi.org/10.4324/9781315276502
37. Medicines & Healthcare products Regulatory Agency, 'Software and Artificial Intelligence (AI) as a Medical Device', GOV.UK. https://www.gov.uk/government/publications/software-and-artificial-intelligence-ai-as-a-medical-device/software-and-artificial-intelligence-ai-as-a-medical-device. Accessed 21 Aug 2023
38. EMA, 'Reflection paper on the use of artificial intelligence in lifecycle medicines', European Medicines Agency. https://www.ema.europa.eu/en/news/reflection-paper-use-artificial-intelligence-lifecycle-medicines. Accessed 21 Aug 2023
39. Proposal for a Directive of the European Parliament and of the Council on liability for defective products Presidency draft compromise proposal (2023). https://data.consilium.europa.eu/doc/document/ST-9676-2023-INIT/en/pdf
40. Regulation (EU) 2017/745 of the European Parliament and of the Council of 5 April 2017 on medical devices, amending Directive 2001/83/EC, Regulation (EC) No 178/2002 and Regulation (EC) No 1223/2009 and repealing Council Directives 90/385/EEC and 93/42/EEC (hereafter Medical Device Regulation)
41. Pearl, J.: Probabilistic reasoning in intelligent systems: networks of plausible inference. Morgan Kaufmann (1988)
42. Halpern, J.Y.: Actual Causality. MIT Press (2019). https://mitpress.mit.edu/9780262537131/actual-causality/. Accessed 01 Aug 2023
43. Tambiama Madiega and European Parliamentary Research Service, 'Artificial intelligence liability directive', PE 739.342 (2023)

44. Nawaz, S.A.: The proposed EU AI liability rules: ease or burden? European Law Blog. https://europeanlawblog.eu/2022/11/07/the-proposed-eu-ai-liability-rules-ease-or-burden/. Accessed 01 Aug 2023
45. BSI and AAMI, 'BS/AAMI 34971:2023 Application of ISO 14971 to machine learning in artificial intelligence – Guide'. https://standardsdevelopment.bsigroup.com/projects/2020-02770#/section. Accessed 23 Apr 2023
46. Onitiu, D.: The limits of explainability & human oversight in the EU commission's proposal for the regulation on AI- a critical approach focusing on medical diagnostic systems. Inf. Commun. Technol. Law **32**(2), 170–188 (2023). https://doi.org/10.1080/13600834.2022.2116354
47. BS EN ISO 14971:2019+A11:2021 Medical devices. Application of risk management to medical devices
48. Kostick-Quenet, K.M., Gerke, S.: AI in the hands of imperfect users. npj Digit. Med. **5**(1) (2022). https://doi.org/10.1038/s41746-022-00737-z
49. Wei, T., Feng, F., Chen, J., Wu, Z., Yi, J., He, X.: Model-agnostic counterfactual reasoning for eliminating popularity bias in recommender system. In: Proceedings of the 27th ACM SIGKDD Conference on Knowledge Discovery & Data Mining, pp. 1791–1800 (2021). https://doi.org/10.1145/3447548.3467289
50. Pfohl, S.R., Duan, T., Ding, D.Y., Shah, N.H.: Counterfactual reasoning for fair clinical risk prediction. In: Proceedings of the 4th Machine Learning for Healthcare Conference, PMLR, pp. 325–358 (2019). https://proceedings.mlr.press/v106/pfohl19a.html. Accessed 01 Aug 2023
51. Braham, M., van Hees, M.: An anatomy of moral responsibility. Mind **121**(483), 601–634 (2012). https://doi.org/10.1093/mind/fzs081
52. Yazdanpanah, V., Dastani, M.: Quantified degrees of group responsibility'. In: Coordination, Organizations, Institutions, and Norms in Agent Systems XI, COIN 2015. LNCS, vol. 9628, pp. 418–436. Springer, Cham (2016). https://doi.org/10.1007/978-3-319-42691-4_23
53. Dastani, M., Yazdanpanah, V.: Responsibility of AI systems. AI Soc. **38**(2), 843–852 (2023). https://doi.org/10.1007/s00146-022-01481-4
54. Dignum, V.: Responsibility and artificial intelligence. In: Dubber, M.D., Pasquale, F., Das, S. (eds.) The Oxford Handbook of Ethics of AI. Oxford University Press (2020). https://doi.org/10.1093/oxfordhb/9780190067397.013.12
55. Yazdanpanah, V., et al.: Different forms of responsibility in multiagent systems: sociotechnical characteristics and requirements. IEEE Internet Comput. **25**(6), 15–22 (2021). https://doi.org/10.1109/MIC.2021.3107334
56. the European Coordination Committee of the Radiological, Electromedical and Healthcare IT Industry (COCIR), 'ARTIFICIAL INTELLIGENCE IN EU MEDICAL DEVICE LEGISLATION (2021). https://www.cocir.org/media-centre/publications/article/cocir-analysis-on-ai-in-medical-device-legislation-september-2020.html. Accessed 20 Apr 2023
57. Middleton, S.E., Letouzé, E., Hossaini, A., Chapman, A.: Trust, regulation, and human-in-the-loop AI: within the European region. Commun. ACM **65**(4), 64–68 (2022)

Limitations of Transparency in Democratising and Regulating Algorithmic Management

Miranda Cross

Oxford Internet Institute, Oxford University, Oxford, UK
miranda.cross@oii.ox.ac.uk

Abstract. This paper discusses the limitations of transparency and explainability requirements for management algorithms as a method of achieving the democratisation of AI governance at work. Using Self-determination theory's conception of basic needs (autonomy, relatedness, competence), I analyse the limitations of transparency as a remedy for the harms that employees experience from algorithmic management, and discuss how failing to address workers' basic needs is op-positional to the goal of democratizing AI in the workplace.

Keywords: Algorithmic management · self-determination theory · future of work

1 Introduction

Modern employers have harnessed the power of machine learning algorithms and increased computing power to automate managerial decisions and increase the productivity of their workers through 'algorithmic management' [1]. These algorithms take in vast amounts of data on workers—collected by surveilling their movements, communications, and outputs—and, through opaque machine learning processing, generate de-cisions that affect everything from pay to termination [2]. Algorithmic management promises better decision-making than human managers, leading to greater efficiency and productivity from workers and higher profits for companies [1]. However, algorithmic management is linked to increases in employee burnout, reports of wrongful termination, and isolation at work [2]. AI regulators tend to argue for transparency in the decision-making processes of algorithms as a remedy for negative effects and to further 'democratise' AI by ensuring that AI is made and governed in line with the will of the majority of the general public. However, does a window into algorithmic man-agement change the experiences of workers who are subject to these decisions and al-low them to exert democratic control over these systems?

In this paper, I seek to investigate the limitations of transparency as a remedy to the harms that algorithmic management systems create for workers and as a method of democratising AI governance in the workplace. First, I present current literature on algorithmic management, and discuss the movement for transparency and democratisation of AI. Utilizing Self-determination Theory's conception of 'basic needs,' I analyse the limitations of transparency as a remedy for employee harms and discuss how failing to address workers' basic needs is oppositional to democratizing AI in at work.

2 Literature Review

2.1 Algorithmic Management

Algorithmic management has been conceived of as a new kind of "digital Taylorism," a 21st-century application of Fredrick Taylor's strategy of 'scientific management' which seeks to control and improve the productivity of workers at any cost [3]. Coined by Lee et. al. [1], the term 'algorithmic management' refers to 'software algorithms that assume managerial functions and surrounding institutional devices that support algorithms in practice' (p. 1). In addition to the delegation of managerial functions, a key feature of algorithmic management is the capture of employee data, 'which fuels the predictive modelling techniques' [2]. Management algorithms automate or inform managerial decisions surrounding hiring, firing, promotions, pay setting, and scheduling across a variety of work sectors. Though extensively used in digital platform work (e.g., rideshare and food delivery platforms), algorithmic management has continued to proliferate across all employment sectors at a rapid rate, including in the managing of full-time employed freight employees, parcel carriers, and warehouse workers [5, 6]. How-ever, it is worth noting that algorithms currently used in full-time employment contexts predominately provide tools to aid human managers, unlike in the gig economy where they are used as direct substitutes for human managers [7].

From the perspective of businesses owners and shareholders, there are a myr-iad of potential benefits associated with the use of algorithmic management including decision-making accuracy, operational efficiency, and improved scalability of business operations [5, 8]. However, from workers' perspectives, the widespread deployment of algorithmic management has had predominately negative consequences, including feel-ings of dissociation and dehumanization and stress from constant surveillance at work [2, 9]. Ajunwa [10] has termed management algorithms as 'the black box at work,' arguing that the proliferation of management algorithms 'simultaneously demands a higher level of transparency from the worker in regard to data collection, while shroud-ing the decision-making in secrecy' (p. 1). She also argues that proliferation of algo-rithms at work presents dangers to the rights and autonomy of workers outside of the immediate work context, because of 'the danger of a 'mission creep' attitude to data collection that allows for pervasive surveillance, contributing to the erosion of both the personhood and autonomy of workers' (p. 1). Proponents of algorithmic management also argue that algorithms can make less biased decisions than human managers, ex-cluding the use of protected characteristics like gender, race, or nationality in decision-making [11]. However, a significant body of research has argued that algorithmic sys-tems do in fact inherit and replicate the biases of algorithm designers, historical HR decisions, and population-level inequalities [3, 8].

2.2 Transparency and Democratisation of AI

The movement towards fairness, accountability, and transparency (FAT) in machine learning (ML), which emphasizes the need for statistical and normative fairness controls, audits/third-party monitoring, and explainable AI/disclosures of AI use to be in-tegrated into the design and use of machine learning, has gained momentum, in part due to several high-profile incidents where ML algorithms were found to be biased against marginalized groups [12, 13]. Transparency, which is the focus of this paper, is often operationalized as a design feature of Explainable AI (XAI) which seeks to de-velop machine learning models that are understandable by humans in terms of data inputs and decision-making outputs [14]. FAT-aware designers of ML algorithms have recently turned their attention towards 'correcting' hiring algorithms to attempt to re-move biases with mixed success—researchers note that the opacity of these algorithms presents a roadblock to evaluating fairness claims and studying their effectiveness in practice [15]. Other applications of algorithmic management, including productivity tracking and algorithmic pay-setting, have not yet been the focus of published fairness assessments because their underlying algorithms are proprietary.

The 'democratisation of AI', a phrase used to refer to a variety of goals including: (1) the democratisation of AI use, (2) the democratisation of AI development, (3) the democratisation of AI profits, and (4) the democratisation of AI governance, is con-cerned with bringing decisions about AI closer to the end users rather than an oligarchy of AI experts and companies [16]. In a workplace context, as workers cannot them-selves 'use' managerial algorithms (only be subject to their outputs) nor develop or profit from them, the democratisation of AI (or algorithmic management, specifically) governance is the only achievable goal. Democratisation of AI governance is concerned with 'ensuring that decisions around questions such as AI usage, development, and profits reflect *the will and preferences of the people being impacted*' (emphasis added)—here, workers—through the reduction of unilateral decision-making and im-plementation of democratic processes [14, pg. 6]. When considering transparency as a method of democratising the governance of AI, the end goal is that affected workers would have increased agency in decision-making about whether to be subject to mana-gerial algorithms, knowledge about which outputs are produced by those algorithms, and the ability to freely opt-out of usage.

Krasnova, in [5], argues that the psychological theory of human motivation called 'Self-determination theory' (SDT) can be used to understand why worker dissatisfac-tion with algorithmic management may be detrimental to firms as well as the workers themselves. SDT 'differentiates between three basic psychological needs: the need for autonomy, competence, and relatedness' (see Sect. 3) [17]. When these needs are satis-fied, people are more likely to be intrinsically motivated, which leads to higher levels of performance, well-being, and satisfaction. Satisfaction of these needs is also crucial to the democratisation of AI governance at work, as they mirror fundamental human rights enshrined in democratic governance and contribute to individual agency in decision-making, which is a key part of participation in a democratic society [18]. Algorithmic management deprives workers of autonomy by stripping them of the abil-ity to make decisions in the workplace and diminishes workers' competence and relat-edness. Ulti-mately, the undermining of basic psychological needs at work may translate into lower performance, reduced motivation, and worsened well-being, as well as thwart the process of democratising algorithmic management.

3 Limitations of Transparency as a Mechanism of Democratisation and Remedy to Harms

Although researchers, including Remus (in [5]), Möhlmann et. al. [4], and Gal et. al. [19], have argued that the opacity of algorithmic management systems is a key problem for designers and employers to solve, simply attending to the problem of transparency does not address the underlying dynamics between workers and employers which make algorithmic management harmful, nor does it meet the needs presented within the SDT framework, which are crucial to participation in democratic AI governance.

3.1 Autonomy

SDT defines the basic psychological need for autonomy as 'individuals' need to act with a sense of ownership of their behaviour and feel psychologically free' and to 'act with a sense of choice and volition' [20]. Algorithmic management is fundamentally at odds with the concept of autonomy because it strips away employee autonomy in mak-ing decisions at work. Transparency in algorithmic management—e.g., workers being aware of what kinds of data are collected, and how decisions are made within algo-rithms— does not solve the problem of loss of autonomy because of the underlying power imbalance at work between employers and employees. Employees cannot freely consent to some types of surveillance at work (e.g., video monitoring, wearables) be-cause the consequences of withholding consent are severe, ranging from 'lower' per-formance, to missed promotions, to potentially losing their employment [6]. Calls for transparency are attempts to correct the information asymmetry between employers and employees. However, the information asymmetry of management algorithms (and their underlying surveillance) is only a surface-level problem. Even when surveillance and algorithmic prompting are transparent and consistently obvious, i.e., with wearable sur-veillance technologies, employees do not have full ability to engage in meaningful re-sistance to this surveillance because doing so would incur the threat of employer re-course, including termination.

The severely limited ability of employees to advocate for increased rights pro-tection in the workplace is well established in other areas of labour law—for example, in the United States, workplaces are expected to adhere to minimum safety standards which are externally enforced by the Occupational Safety and Health Administration [21]. Workers are not expected to constantly advocate to their employers for protections against, for example, toxic chemical exposure, because doing so would place them at constant risk of job loss. Minimum safety standards, along with other protections like minimum wage, child labour standards, and overtime pay protect workers from dis-crimination in the labour marketplace for refusing to consent to work in unsafe or un-derpaid working

conditions. The same minimum standard does not yet exist for algo-rithmic surveillance in the EU[1] or US— thus, transparency standards do not eliminate
the risk of job loss to workers who seek autonomy in decision making or seek to resist surveillance. Loss of autonomy is in direct conflict with democratisation of algorithmic management because freedom of choice (as opposed to coercion) is a fundamental tenet of democratic participation. Here, transparency does not contribute to democratisation of AI at work because neither notification of being subject to algorithmic decision-making nor understanding the outputs of those algorithms can give workers the freedom to choose not to be algorithmically managed.

3.2 Competence

The ability to feel a 'sense of mastery over the environment and to develop new skills' is referred to in SDT literature as 'competence,' the second of the core psychological needs [20]. Algorithmic management and underlying workplace surveillance limit all workers' ability to develop competence because it interrupts traditional skill-building pathways. Managers, for example, are deprived of opportunities to build 'tacit knowledge' through their experience of dealing with employees and instead become reliant on algorithms to inform or make decisions [3]. Employees are subject to opaque metrics set by algorithms that may not be intrinsically tied to performance but rather seek to optimize productivity and profitability for the company. Algorithmic transpar-ency can reduce the opacity of these metrics, but it cannot replace individualized man-agerial feedback about how to build skills. Algorithmic management, as discussed above, also disrupts traditional markers of job performance, including promotions and wage setting. Even if workers know that a wage is being set due to demand, transpar-ency around algorithmic wage setting does not increase feelings of competence— the algorithm maximizes profitability, rather than rewarding quality work.

In addition, workers cannot feel competent at work if they have no redress for incorrectly-made algorithmic managerial decisions— the precarity of being subject to algorithmic decisions which may not be based on accurate data or may utilize flawed decision-making undermines any ability to feel 'mastery' over one's work environ-ment. This phenomenon is particularly evident in the case of Amazon warehouse work-ers and drivers, who have no human manager to report issues to. An example of flawed algorith-mic decision-making is Amazon Flex's performance rating for drivers not con-sidering traffic delays [23]. When workers are fired for 'poor performance' or factors out of their control, there is no traditional mechanism of redress; or, as one Amazon worker put it: "whenever there's an issue, there's no support . . . It's you against the machine, so you don't even try" [23]. There is no incentive to develop competence at work if workers know they can be penalized, without recourse, due to a faulty algo-rithm. Lack of competency is in direct opposition to democratisation of AI because democratisation is concerned with the sharing of knowledge—the inability to develop competency at work, and particularly competency in understanding the algorithms which govern the

---

[1] Though some individual uses of algorithmic management tactics have been ruled illegal in the EU— for example, in the Italian Data Protection Authorities cases which ruled against the "rider ranking" algorithms used by Foodinho and Deliveroo— these have been ruled illegal.

workplace, makes democratisation challenging because of a lack of knowledge which limits participation.

3.3 Relatedness

The basic need for relatedness in SDT 'is satisfied when people see themselves as a member of a group, experience a sense of communion, and develop close relations' [20]. Both algorithmic management, which partially or fully removes human managers from managerial decisions, as well as its underlying workplace surveillance represent key barriers to developing close relationships in the workplace. Although this is already evident in the platform economy, where workers are more likely to be 'independent' and have few to no workplace connections, increased use of algorithmic management in traditional workplaces also risks disrupting employees' ability to form connections with other employees. Transparency in algorithmic management does not solve the problem of disconnection—even in instances in the gig economy where workers are fully aware that they are being managed by an algorithm, they report intense feelings of 'dehumanization' and isolation at work [9]. In traditional work settings, algorithms disrupt the traditional feedback cycle between managers and employees, leading to a loss of relatedness even if algorithms are clear about the data they collect and how it is analysed. For example, in promotions and layoff contexts, managers may not be able to give feedback about employees' job performance if decisions about promotions and layoffs are based on the recommendations of an algorithm[2]. When using algorithms to schedule shift work, managers cannot override scheduling algorithms to accommodate for emergencies or other situations that arise outside of work, depriving employees of a crucial point of connection and empathy.

The digital and physical surveillance which underpin algorithmic management also impede the development of workplace relationships, leading to a loss of related-ness. Employees subject to monitoring of their chats, emails, and even spoken conver-sations must constantly monitor their self-expression to avoid running afoul of com-pany policies, making connection-building with fellow employees much harder. This has particularly disastrous consequences for collective action among employees at work— worker surveillance has long been used to stymie potential collective action (including union organizing) among workers, but increasingly invasive surveillance, even if transparently conducted, makes collective action almost impossible to organize. For example, as a way of monitoring employee communications, Amazon requires the use of a proprietary chat app, which bans words including 'union' and 'grievance' [24]. Because the data collected by such surveillance is necessary for algorithmic manage-ment to make decisions, it is unlikely that transparency—whether regarding the data collected or algorithmic decision-making—would address the underlying negative con-sequences of surveillance on workplace relationships.

Making algorithmic management practices transparent, particularly in a promotion/layoff context, poses specific problems to building relatedness— as Krasnova (in [5]) argues, 'visibility of one's performance and a related scoring system may trigger

---

[2] This is more applicable in a US context, as Article 22 of the GDPR/UK GDPR give European and British citizens the right not to be subject to solely automated decisions.

competitive behaviours among workers, which over time may undermine collaboration and workplace climate' (p. 828). This conclusion is supported by evidence from Levy [25] wherein truck drivers who were subjected to electronic monitoring systems reported increased competitiveness with other drivers as well as a breakdown of tradi-tional information-sharing pathways between drivers. Increased competitiveness in the workplace is antithetical to workers' feelings of relatedness—when artificially en-hanced by transparency in algorithms, increased competition poses a significant threat to the basic need for relatedness at work. Relatedness is fundamentally tied to democ-ratisation because of the people-centric process of democracy which AI democratisa-tion attempts to replicate; however, without the ability to feel related to other citizens (or fellow workers), these processes break down due to a lack of trust.

4 Discussion

Workers' fulfilment of the three basic psychological needs of autonomy, competence, and relatedness in the SDT framework is crucial to their development of intrinsic mo-tivation, leading to better performance and greater well-being at work. Algorithmic management, in its myriad forms, presents a substantial risk to employees' ability to fulfil these needs, thus creating undesirable and, in some cases, unsafe work environ-ments. Transparency requirements have thus far been a focus of the policy remedies which seek to curb the harms of algorithmic management; however, they are not a pan-acea. The opacity of algorithms is only one problem—in sectors where there is some degree of algorithmic transparency, including in the platform economy, employees' ability to fulfil their basic psychological needs is still inhibited by the human discon-nection that algorithms create as well as the lack of redress for unfair decisions.

Proposed legislative and regulatory governance mechanisms for algorithms at work (most notably the EU AI Act) focus heavily on transparency as a mechanism of accountability and regulation for algorithmic management. However, as discussed above, transparency-based regulations fail to meaningfully address: (1) the underlying power imbalances implicated in employer surveillance of employees, (2) the negative impacts that algorithmic management has on employees, and (3) the process of democ-ratisation of AI at work. These would be better addressed by regulations that include, in addition to transparency, accountability mechanisms for employers which provide redress for incorrectly made algorithmic decisions, and a focus on the minimization of worker data collection and use in algorithmic management [26].

Democratisation at work—both inside and outside of the context of algorithmic man-agement—necessitates the balancing of power between employees and management, which is achieved through mechanisms of increased employee decision-making. Future regulation should look beyond transparency and towards accountability mechanisms and other structural fixes for the underlying problems with algorithmic management which disrupt employees' ability to foster autonomy, competence, and relatedness at work. Only then will regulation achieve the goal of democratising the governance of AI at work.

References

1. Lee, M.K., Kusbit, D., Metsky, E., Dabbish, L.: Working with machines: the impact of algorithmic and data-driven management on human workers. In: Proceedings of the 33rd Annual ACM Conference on Human Factors in Computing Systems, pp. 1603–1612. Association for Computing Machinery, New York, NY, USA (2015). https://doi.org/10.1145/2702123.2702548
2. Jarrahi, M.H., Newlands, G., Lee, M.K., Wolf, C.T., Kinder, E., Sutherland, W.: Algorithmic management in a work context. Big Data Soc. **8**, 1–14 (2021). https://doi.org/10.1177/20539517211020332
3. Brown, P., Lauder, H., Ashton, D.: Digital Taylorism. In: Brown, P., Lauder, H., Ashton, D. (eds.) The Global Auction: The Broken Promises of Education, Jobs, and Incomes. Oxford University Press (2010). https://doi.org/10.1093/ac-prof:oso/9780199731688.003.0016
4. Möhlmann, M., Zalmanson, L., Henfridsson, O., Gregory, R.W.: Algorithmic Management of Work on Online Labor Platforms: When Matching Meets Control. MIS Q. **45**, 1999–2022 (2021). https://doi.org/10.25300/MISQ/2021/15333
5. Benlian, A., et al.: Algorithmic Management. Bus. Inf. Syst. Eng. **64**, 825–839 (2022). https://doi.org/10.1007/s12599-022-00764-w
6. Mateescu, A., Nguyen, A.: Algorithmic Management in the Workplace. Data & Society, United States (2019). https://datasociety.net/wp-content/up-loads/2019/02/DS_Algorithmic_Management_Explainer.pdf
7. Wiener, M., Cram, W., Benlian, A.: Algorithmic control and gig workers: a legitimacy perspective of Uber drivers. Eur. J. Inf. Syst., 1–23 (2021). https://doi.org/10.1080/0960085X.2021.1977729
8. Kellogg, K.C., Valentine, M.A., Christin, A.: Algorithms at work: the new contested terrain of control. Acad. Manag. Ann. **14**, 366–410 (2020). https://doi.org/10.5465/an-nals.2018.0174
9. Möhlmann, M., Henfridsson, O.: What people hate about being managed by algorithms, according to a study of Uber drivers (2019). https://hbr.org/2019/08/what-people-hate-about-be-ing-managed-by-algorithms-according-to-a-study-of-uber-drivers
10. Ajunwa, I.: The 'black box' at work. Big Data Soc. **7**, 1–6 (2020). https://doi.org/10.1177/2053951720938093
11. Sonderling, K.E.: How People Analytics Can Prevent Algorithmic Bias. https://www.ih-rim.org/2021/12/how-people-analytics-can-prevent-algorithmic-bias-by-commissioner-keith-e-sonderling/. Accessed 17 Apr 2023
12. Chouldechova, A.: Fair prediction with disparate impact: a study of bias in recidivism prediction instruments. Big Data. **5**, 153–163 (2017). https://doi.org/10.1089/big.2016.0047
13. Kelly-Lyth, A.: Challenging biased hiring algorithms. Oxf. J. Leg. Stud. **41**, 899–928 (2021). https://doi.org/10.1093/ojls/gqab006
14. Ehsan, U., Liao, Q.V., Muller, M., Riedl, M.O., Weisz, J.D.: Expanding explainability: towards social transparency in AI systems. In: Proceedings of the 2021 CHI Conference on Human Factors in Computing Systems, pp. 1–19. Association for Computing Machinery, New York, NY, USA (2021). https://doi.org/10.1145/3411764.3445188
15. Raghavan, M., Barocas, S., Kleinberg, J., Levy, K.: Mitigating bias in algorithmic hiring: evaluating claims and practices. In: Proceedings of the 2020 Conference on Fairness, Accountability, and Transparency, pp. 469–481. Association for Computing Machinery, New York, NY, USA (2020). https://doi.org/10.1145/3351095.3372828
16. Seger, E., Ovadya, A., Garfinkel, B., Siddarth, D., Dafoe, A.: Democratising AI: Multiple Meanings, Goals, and Methods. arXiv Preprint (2023). https://doi.org/10.48550/arXiv.2303.12642

17. Deci, E., Olafsen, A., Ryan, R.: Self-determination theory in work organizations: the state of a science. Annu. Rev. Organ. Psychol. Organ. Behav. **4** (2017). https://doi.org/10.1146/annurev-orgpsych-032516-113108
18. Boyte, H.C.: Reframing democracy: governance, civic agency, and politics. Public Adm. Rev. **65**, 536–546 (2005)
19. Gal, U., Jensen, T.B., Stein, M.-K.: Breaking the vicious cycle of algorithmic management: a virtue ethics approach to people analytics. Inf. Organ. **30**, 100301 (2020). https://doi.org/10.1016/j.infoandorg.2020.100301
20. Van den Broeck, A., Ferris, D.L., Chang, C.-H., Rosen, C.C.: A review of self-determination theory's basic psychological needs at work. J. Manag. **42**, 1195–1229 (2016). https://doi.org/10.1177/0149206316632058
21. Occupational Safety and Health Administration: OSHA Worker Rights and Protections, https://www.osha.gov/workers. Accessed 17 Apr 2023
22. Lomas, N.: Italian court rules against 'discriminatory' Deliveroo rider-ranking algorithm (2021). https://techcrunch.com/2021/01/04/italian-court-rules-against-discriminatory-deliveroo-rider-ranking-algorithm/
23. Kaur, D.: At Amazon Flex, it's the driver vs the algorithm. https://techhq.com/2021/07/at-amazon-flex-its-the-driver-vs-the-algorithm/. Accessed 17 Apr 2023
24. Klipperstein, K.: Leaked: New Amazon Worker Chat App Would Ban Words Like 'Union,' 'Restrooms,' 'Pay Raise,' and 'Plantation (2022). https://theintercept.com/2022/04/04/amazon-un-ion-living-wage-restrooms-chat-app/
25. Levy, K.: The contexts of control: information, power, and truck-driving work. Inf. Soc. **31**, 160–174 (2015). https://doi.org/10.1080/01972243.2015.998105
26. Dubal, V.: On algorithmic wage discrimination. SSRN (2023). https://doi.org/10.2139/ssrn.4331080

Breaking the Filtered Lens: A Feminist Examination of Beauty Ideals in Augmented Reality Filters

Mariana P. Castillo-Hermosilla(✉) , Hedye Tayebi-Jazayeri ,
and Victoria N. Williams

Universität Osnabrück, Osnabrück, Germany
{mcastilloher,htayebijazay,vwilliams}@uni-osnabrueck.de

Abstract. This paper is a work-in-progress theoretical discussion in the field of Ethics of AI. Drawing from a feminist theoretical perspective, we discuss the political and personal implications of beauty ideals perpetuated by AI technologies such as Augmented Reality (AR) filters or face filters, specifically beauty filters. Most research focuses on the impact of social media on individuals, particularly young women, or has analyzed the impact of AR filters in the context of social media. Therefore, we add a structural analysis to the debate by challenging the notion that conformity to beauty ideals is a personal matter, while stressing the necessity to analyze beauty filters as a phenomenon that extends its influence beyond social media. We explore how beauty filters emerge as both products and perpetrators of unrealistic ideals about feminine bodies and contribute to the ongoing objectification of women, exacerbating social pressure and hyper-fixation on appearance. We also reflect, from a critical phenomenological approach, on how socially constructed habits of female bodies in a patriarchal society are embodied and impact girls' and women's autonomy and freedom. We suggest a critical examination of daily-use AI technologies such as beauty filters, which influence mental well-being, exemplified by phenomena like *Snapchat dysmorphia*. In this light of misogynistic AI, we advocate for a heightened awareness of women's issues within the discourse on AI development.

Keywords: Face filters · Beauty filters · Artificial intelligence · Feminism · Beauty ideals · Ethics

1 Introduction

Social media platforms have become an integral part of daily life. In the last decade, platforms such as *Instagram, Snapchat,* or *TikTok* have introduced face filters, automatic photo editing tools that use artificial intelligence and computer vision [17] designed to achieve "hyper realistic" facial remodeling [1, 10]. Beauty filters are filters that remodel facial features to conform to conventional "beauty," i.e., smooth skin, slim face, small nose, large eyes, and lips, etc., their use is particularly popular among young women

[17]. For example, the *Bold Glamour* filter on TikTok has been used in over 2.8 million videos as of March 2023 [10]. Social media platforms rely heavily on AR technologies, which are referenced as an efficient marketing tool [17].

Many researchers in the field of mental health warn of the negative consequences such tools can bring [16, 18]. Beauty filter technologies have taken the normalization of beauty to a new level by enabling users to conform to media-driven beauty ideals with just a few clicks. This underscores contemporary challenges, including visual overload and pressure to project a perfect life, impacting self-esteem and personal agency [6]. Previous research on filters and social media has revealed that comparing one's unfiltered self with an idealized-filtered self, can cultivate self-insecurities and discrepancies, possibly leading to poor mental health and a desire for cosmetic procedures [5, 8]. Beauty filters contribute to self-referential dissatisfaction [e.g., 2, 5], intensifying the pressure for self-optimization [2, 5, 6], which is especially concerning for young adolescents who are a vulnerable group for negative self-image [2, 5]. The resulting self-image disparity between the real and ideal appearance has even been associated with eating disorders [14]. Furthermore, rates of anxiety and depression in young people have risen by 70% in the past 25 years, with women facing a higher risk of entering a negative feedback loop that exacerbates depression and body surveillance, potentially leading to body dysmorphic disorder (BDD) [11]. BDD affects 1.7% to 2.9% of the general population [9] highlighting the need for education and prevention due to its strong association with filter use [2, 5]. The phenomenon known as *Snapchat Dysmorphia*, coined by cosmetic doctor Tijion Esho, describes the shift from wanting to emulate celebrities to one's filtered self. This has also raised concerns among cosmetic doctors and psychologists who believe it poses significant risks to mental health [5, 8]. These findings underscore the need to approach new technologies with differentiated perspectives. They stress beauty filters' influence on self-image and beauty norms, particularly in women, motivating the subsequent discussion.

The increasing popularity of beauty filters in our capitalist society, which is centered on profit-driven social media companies, warrants closer examination. However, most research focuses on the impact of social media on individuals, particularly young women. Furthermore, there has been little research on the impact of AR filters beyond social media. We would like to add a structural analysis to the debate by challenging the notion that conformity to beauty ideals is exclusively a personal matter and highlighting the interplay between social structures and subjectivity. Our goal is to illuminate the complex misogyny that underlies the rise of beauty filters and reinforces a culture that exploits "beauty" as a means to oppress women. This raises the need to discuss beauty ideals on face filters beyond the individual level and to understand their psychological impact as a consequence of the same context. Nevertheless, given the ubiquity of the negative consequences for mental health regarding the use of beauty filters, we propose to further adopt a phenomenological perspective to bridge the gap between a political critique and an understanding of how given structures are enacted and embodied more concretely. We aim to inspire a theoretical exploration of the far-reaching political and personal consequences of beauty ideals in the digital age, laying a foundation for in-depth future discussions encompassing practices, regulations, etc. In light of the presence of misogynistic AI, e.g., as seen in AI "girlfriends", sex robots or new phenomena such as

AI influencers, we advocate for a heightened awareness of women's issues within the discourse on AI development.

2 Beauty Filters: How They Work and Exert Implicit Power

Face filters, or Augmented Reality (AR) filters, are automated tools that employ AI and computer vision to detect and modify facial features [17]. Originating from *Lookserly*, they introduced the real-time customization of facial features during video chats or photo/video capture [1]. Today, various platforms incorporate them, so they are found not only on social media but on specialized photo editing apps like *Fotor*, *BeautyPlus*, *YouCam Perfect*, etc.[1] This technology analyses pixel colours in images to identify facial features and applies pixel-level transformations through image processing [1, 17]. It creates a precise facial model, adjusting predefined points to follow facial movements, allowing for various modifications [1]. Trained on large face image datasets, these models recognize different faces and adapt masks based on user features and filter functions.[2] While the functionality may be politically relevant, it won't be discussed further. Instead, given our focus on beauty ideals,[3] we suggest the following findings exemplify the exertion of norms.

Unveiling Racist Inclinations in Beauty Filters. Beauty filters have been criticized for perpetuating racism and Eurocentric beauty ideals by lightening skin or reducing nose size [15, 17]. Riccio and Oliver's examination of the AR-based selfie beautification algorithm, *Beautyverse*, has revealed that beauty filters tend to make individuals of all ethnicities appear whiter. They found increased misclassification of beautified faces as white by race classification algorithms, *DeepFace* and *FairFace*, even in non-beautified cases. This highlights how Beautyverse homogenizes facial aesthetics, making them conform to a 'white beauty' standard [15]. The implied racial dimension adds complexity for non-white women, heightening the challenges related to beauty ideals. Emphasizing the uniqueness of 'white beauty,' it is vital to critically examine the interplay between technology, beauty standards, and race, unveiling intricate entanglements that amplify women's experiences.[4]

---

[1] A Google search yields various articles recommending photo editing apps, like the one found at: https://www.fotor.com/blog/face-filter-app/

[2] We provide this concise technology clarification; more extensive technical details are available in the references, which we won't delve into as it doesn't enhance our argument.

[3] Due to space constraints, we cannot adequately delve into the intricate relationship between beauty ideals and racism, but we do aim at implicitly connecting them in virtue of discussing racism in the first place.

[4] We follow Crenshaw's notion of intersectionality (1989), revealing how various forms of discrimination, including race, gender, and social class create cumulative disadvantages that require a holistic perspective.

3 Beauty Ideals and Beauty Filters: A Feminist and Phenomenological Perspective

In today's digital capitalist[5] era, technology and consumer markets magnify stereotypes in mass media. Beauty filters are only *seemingly* harmless as they do carry a deeper weight by perpetuating beauty ideals. They are contextualized within a broader, politically relevant, socio-technological framework that affects social norms, self-perception and questions of power [3]. Thereby, beauty filters extend their influence beyond technology, becoming a tool to sustain power relations. Understanding beauty ideals in a structural and political context is essential, given the historical objectification and control of the female body [3]. This subjugation has transformed to a broad culture of over-sexualization and commodification [3, 22]. Such socio-historically sensitive considerations become integral in understanding how technologies such as beauty filters continue to objectify the female body. Technological progress, particularly in cameras, videos, and visually focused social media platforms like Instagram, has amplified the emphasis on women's visual appearance [22]. Beauty filters, as a product and enhancer of these mechanisms, vividly reflect these societal trends. By enabling users to conform to media-driven beauty ideals, they actively engage in shaping their digital looks. Much like traditional advertising historically influenced consumer behavior and beauty ideals [22], the widespread use of beauty filters normalizes certain looks, influencing our perceptions through exposure and our interaction with the technology (cf. *Human-aided AI*).

To explore the interplay between beauty ideals and their effect on women's agency, the notion of the *lived body* may help to understand how people embody, re-act, and reproduce beauty ideals [23]: this contrasts with the notion of subjectivity as the sole principle of all appearances, as meaning and potential actions emerge through direct engagement with the world. This situatedness stresses the influence of culture on the construction of meaning and identity formation since the lived body is enculturated [23]. Women acquire habits of feminine body comportment that come from their situation in a patriarchal society: as they learn to live in accordance with femininity mandates, "*the more she takes herself to be fragile and immobile and the more she actively enacts her own body inhibition*" [23]. These mandates are relationally constituted and exert tangible influence on individuals' social standing, self-perception, and practical agency. Such influence and objectification of feminine bodies changes the way in which women experience the power over their own bodies and movements, impacting their confidence and amplifying their self-consciousness and self-objectivization: their body is an object that is seen and acted upon by others, so they worry about how it looks, and they shape, model, and decorate it [23].

Beauty Filters follow mostly Eurocentric beauty standards [15], perpetuating the idea of them as *the* features considered beautiful by society. Currently, the filters offered by the App/Play Store do not just edit the photos; some offer to rate, track, or even "improve" women's appearance. "Rating" apps, e.g., *FaceRate* analyze the user's pictures and rate "how attractive" or "how symmetrical" one's face is. This amplifies the appearance pressure, even when young girls and women know how unrealistic these ideals and

---

[5] We use "digital" capitalism to explore the effects of widespread digital technology adoption, aligning with our specific focus on beauty ideals and capitalism in the context of beauty filters.

rates are [6]. Furthermore, due to the HD technology offered by cameras and the possibility to change one's features while looking at them by using filters, this generation has developed new visual skills to measure and evaluate their own appearance, practicing nano-surveillance of their bodies and faces [6]. Filters such as *Fix-Me*, banned by Facebook in 2019 as they "*fuel insecurity amongst young Brits*" and appear to promote plastic surgery [19], play to this "forensic" view: the structural mechanisms of the filters shape the attitudes towards oneself and others. For instance, the filters can contribute to a hyper-focus on perfection by addressing perceived "faults", e.g., blemishes and wrinkles, thus promoting a fixation on such aspects. While this level of hyper-fixation might not be as pronounced in general platform use, the reach of surveillance is reconfigured. The tools, on the one hand, foster a subjectivity beyond traditional self-care practices promoting a culture of meticulous self-examination that enforces normative visual appearance [4], and on the other hand, it changes the way in which the new generations experience and even perceive their bodies, furthering self-objectification.

Conforming to beauty standards is often presented as a personal choice. However, following the previous reflection regarding how social norms are embodied and observing how women are evaluated primarily based on their sexual attractiveness, we raise questions about the true freedom of these choices and their impact on socio-economic opportunities. This demand extends beyond adhering to beauty ideals; it requires aligning internal desires, and reinforcing subordination through *self-surveillance* [3] and arguably, *self-objectification*. Elias and Gill, in their analysis of beauty apps, aptly point out that these demands have "*intensified rather than diminished, albeit wrapped in discourses that highlight pleasure, choice, agency, confidence and pleasing oneself, obscuring to which aesthetic labour on the body is normatively demanded*" [4]. Beauty practices have been also associated with the idea of self-esteem and general well-being, reframing "beauty adherence" as a sign of "good mental health" [20], making it difficult to detach oneself from these ideals. Thus, women use filters to "improve" their appearance, reinforcing certain norms in both virtual and real spaces. The beauty pressure increases body dissatisfaction, with women reporting not feeling comfortable leaving their house without makeup [e.g., 6] and normalizes cosmetic procedures [e.g., 2]. These procedures not only bring with them the general medical risks but also homogenize women's bodies as they aim to shape it to conform to existing beauty ideals, thereby reinforcing "*A* correct way" female bodies should look like.

In summary, the constant exposure to and interaction with beauty filters has changed the ways in which women perceive themselves and others, subsequently influencing their behaviors. These changes reinforce the existing ideals about beauty, homogenizing bodies and excluding diverse or non-normative bodies. The established subjectivity, i.e., desire to use beauty filters, relevate the question of women's autonomy.

Thereby we call for an analysis of the introduction of beauty filters into the societal context just presented, with recognition of its relevance in light of the fight for women's liberation. This perspective is also crucial for analyzing the psychological repercussions of AR filters and the emergence of new phenomena such as Snapchat dysmorphia. Inspired by de Ugarte's notion of *political self-esteem* [20], we must deprivatize our emotional experience and self-understanding, and understand them as deeply rooted and affected by gender roles and beauty ideals, that are amplified by these technologies.

4 Conclusion

Existing research mainly addresses mental health effects of Beauty Filters, or racial biases in them [e.g., 2, 5, 6, 15]. Yet, the link between replicating beauty ideals and mental health has been poorly explored. Within this context, the lived body gains significance in understanding how individuals embody these ideals and norms [23], contributing to the deprivatization of women's problematics, but also highlighting the role of beauty filters in amplifying this pressure. Beauty filters heighten self-awareness of appearance influencing self-perception, normalizing cosmetic procedures, and eroding diversity. Our discussion sheds light on the mental health effects of these technologies by considering how societal structures generate discomfort. On the other hand, we keep the discussion open about how others, e.g., gender non-conforming people, may also be impacted by these technologies.

Socially constructed habits profoundly shape the agency of women [23]. In our digital capitalist world, technologies like beauty filters gain political relevance, as they emerge as both products and perpetrators of these dynamics, highlighting the ties between beauty, power, subjectivity [3]. Despite a semblance of agency in their usage, a more profound analysis exposes that these technologies take on a notably misogynistic nature within the cultural context, i.e., patriarchal capitalism. As a result, these tools not only sustain prevailing beauty pressures but amplify them, restricting women's freedom due to the demand to adhere to entrenched beauty norms. Future research may further look into AR technologies' use of "gamification" and "social exploit" to collect data [13]. The ethics of biometric data generation largely focuses on coerced surveillance, e.g., by state or corporate entities like *Google* [4]. By highlighting how "*biometric rationality runs through contemporary beauty culture*" [4], one may problematize 'voluntary' self-surveillance stemming from the described mechanisms. To grasp how AR filters replicate beauty ideals, one may further examine their embedded power dynamics and economic motives. "Biases" are tied to established practices within organizations where datasets and models are burgeoned. Within the capitalist context of their development, social media platforms and AR filter apps capitalize on people's natural inclination for social participation to develop, train, and profit. Overall, this political outlook aims to emphasize the need to regard women's issues in discussions surrounding AI.

References

1. Bansal, M.: The technology behind Face Filters - IEEE Women in Engineering, VIT – Medium. https://medium.com/ieee-women-in-engineering-vit/technology-behind-face-filters-98e234b9fc33. Accessed 14 June 2023
2. Beos, N., Kemps, E., Prichard, I.: Photo manipulation as a predictor of facial dissatisfaction and cosmetic procedure attitudes. Body Image **39**, 194–201 (2021)
3. Dimulescu, V.: Contemporary representations of the female body: consumerism and the normative discourse of beauty. https://www.semanticscholar.org/paper/Contemporary-Representations-of-the-Female-Body%3A-of-Dimulescu/1b6ba62257b7dfc56ce73dc160534ce2d36e54b4 (2015)
4. Elias, A.S., Gill, R.: Beauty surveillance: the digital self-monitoring cultures of neoliberalism. Eur. J. Cult. Stud. **21**(1), 59–77 (2017)
5. Eshiet, J.: "Real me versus social media me": Filters, Snapchat Dysmorphia, and beauty perceptions among young women. Electronic Theses, Projects, and Dissertations. 1101 (2020). https://scholarworks.lib.csusb.edu/etd/110

6. Gill, R.: Changing the Perfect Picture: Smartphones, Social Media and Appearance Pressures. University of London, City (2021)
7. Higgins, E.T.: Self-discrepancy: a theory relating self and affect. Psychol. Rev. **94**(3), 319–340 (1987)
8. Hunt, E.: Faking it: how selfie dysmorphia is driving people to seek surgery. The Guardian, https://www.theguardian.com/lifeandstyle/2019/jan/23/faking-it-how-selfie-dysmorphia-is-driving-people-to-seek-surgery. Accessed 16 Aug 2023
9. International OCD Foundation: Prevalence of BDD. https://bdd.iocdf.org/professionals/prevalence/. Accessed 17 Aug 2023
10. Koh, R.: TikTok's new beauty filter is so realistic that people can't tell when it's being used. Psychologists, aestheticians, and neurologists are all concerned, Insider, https://www.insider.com/experts-warn-about-tiktoks-creepily-realistic-viral-beauty-filter-2023-3. Accessed 20 Aug 2023
11. Lamp, S.J., Cugle, A., Silverman, A.L., Thomas, M.T., Liss, M., Erchull, M.J.: Picture perfect: the relationship between selfie behaviors, self-objectification, and depressive symptoms. Sex Roles **81**, 704–712 (2019)
12. McGrath, L.R., Oey, L., McDonald, S., Berle, D., Wootton, B.M.: Prevalence of body dysmorphic disorder: a systematic review and meta-analysis. Body Image **46**, 202–211 (2023)
13. Mühlhoff, R.: Human-aided artificial intelligence: or, how to run large computations in human brains? Toward a media sociology of machine learning. New Media Soc. **22**(10), 1868–1884 (2019)
14. Quittkat, H.L., Hartmann, A.S., Düsing, R., Buhlmann, U., Vocks, S.: Body dissatisfaction, importance of appearance, and body appreciation in men and women over the lifespan. Front. Psych. **10**, 1–12 (2019)
15. Riccio, P., Oliver, N.: Racial bias in the beautyverse: evaluation of augmented-reality beauty filters. In: Karlinsky, L., Michaeli, T., Nishino, K. (eds.) Computer Vision – ECCV 2022 Workshops, LNCS, vol. 13803, pp. 714–721. Springer, Cham Tel Aviv (2023). https://doi.org/10.1007/978-3-031-25066-8_43
16. Royal Society of Public Health: Status of Mind: Social media and young people's mental health. Young Health Movement, London (2017). Used once
17. Ryan-Mosley, T.: Beauty filters are changing the way young girls see themselves, MIT Technology Review. https://www.technologyreview.com/2021/04/02/1021635/beauty-filters-young-girls-augmented-reality-social-media/. Accessed 14 2023/06/14
18. Sampasa-Kanyinga, H., Lewis, R.F.: Frequent use of social networking sites is associated with poor psychological functioning among children and adolescents. Cyberpsychol. Behav. Soc. Netw. **18**(7), 380–385 (2015). Used once
19. The Beauty Gurus: Instagram 'fuelling mental health crises'with "Fix Me" plastic surgery filter. https://www.thebeautygurus.com/blog-article/instagram-fuelling-mental-health-crisis-with-fix-me-plastic-surgery-filter/. Accessed 16 Aug 2023
20. Ugarte, N.: La dictadura del amor propio: sobre positivismo tóxico, autoestima y salud mental. Penguin Random House Grupo Editorial, Santiago de Chile (2022)
21. ScienceDaily: It's not if, but how people use social media that impacts their well-being: Passively scrolling through posts may not result in feelings of happiness. www.sciencedaily.com/releases/2020/11/201102110030.htm. Accessed 11 May 2023
22. Wolf, N.: The Beauty Myth: How Images of Beauty are used Against Women. Random House, New York (2013)
23. Young, I.M.: On female body experience: "Throwing Like a Girl" and Other Essays (C. Calhoun, Ed.). Oxford University Press, New York (2005)
24. Zhao, S., Zappavigna, M.: The interplay of (semiotic) technologies and genre: the case of the selfie. Soc. Semiot. **28**(5), 665–682 (2018)

Data Privacy and Technology Ethics

Your Body Should Not Belong to the Internet: Online Bodily Integrity in the World of Deepfake Pornography

Lyndsey Scott[✉]

University of Cambridge, Cambridge, England
ls2023@cam.ac.uk

Abstract. *To what extent do we own ourselves and our bodies when our privacy is violated online?* In this paper, I argue that distancing the online and physical self risks diminishing the impact of privacy violations. I frame the concept of informational privacy in the scope of generated deepfake content online, focusing on pornographic imagery generated unknowingly without the participant's consent. I propose that informational privacy does not adequately cover the violation of these attacks and offer that ideas around bodily integrity should be extended to attacks on the online self and body, a new form of *online bodily integrity*.

Keywords: deepfake pornography · informational privacy · bodily autonomy · online behaviour · generative AI

1 Introduction

The relationship between the self and the body has been a long-considered debate [21]. With the increased digitisation of today's modern world, a complication one has to contend with is a third aspect of these two concepts; the online self or "digital self" [34]. When considering the many forms of privacy and autonomy related to the self, the expectations for an online self can be distinctly different. In this paper, I make the argument that we should consider privacy attacks on the online self with a more tightly woven relationship to our physical self. Considering existing concepts around informational privacy and the online self and body, I propose that using deepfake technology to generate non-consensual pornography is at a unique intersection of these two areas. A key concept in bodily integrity is the intersection of political systems and societal structures influencing the degree of autonomy to which one can control their own body. In particular, I build on the proposal that this act should qualify as an attack on a newer form of *online bodily integrity*. I put forward that distancing the online self from the physical self undermines the reality faced by victims of these attacks and makes space for both a lack of empathy from observers and consequences for the perpetrator. Subsequently, improper informational privacy concepts for these attacks have the potential to slow down effective legislation and action for victims, especially if they are the only lens we use to consider these online crimes.

Deepfake revenge porn is a particularly potent and relevant instance of a privacy violation to consider with this argument. Developments in generative artificial intelligence (AI) have made it significantly easier for non-consensual imagery to be created without significant effort, engineering skill, or photo manipulative ability [10]. Although image manipulation has occurred since the 19th century [29], the ease of access that deepfake technology enables provides a new frontier to contend with. Online, sexual imagery can be weaponised to target and humiliate victims, and these attacks disproportionately affect women [10].

In 2019, it was estimated 96% of deepfake imagery online is pornography [9]. In 2021, Sensity AI found that non-consensual pornography made up 90–95% of deepfake content online [13]. In recent years, highly publicised breakthroughs in generative AI such as DALL-E and Stable Diffusion have been released for public consumption. Although the newer, most used publicly available models such as DALL-E have safeguards against generating pornographic content, it is a natural assumption that advanced ML techniques with pornographic imagery will be developed in their wake in the near future. Stable Diffusion has already been manipulated to do so [28]. A higher rate of deepfake revenge porn attacks with more even convincing imagery is a considerable threat to privacy, especially combined with doubt over the effectiveness of legislation worldwide on the specific act of deepfakes. For those at higher risk of sexual violence, the online landscape appears growingly hostile.

To situate this argument, I first introduce deepfake technology and its use of it for non-consensual pornographic imagery. Using the experiences of victims, I argue that this form of privacy violation is distinct and requires a new language. In the second part, I discuss how modern interpretations of the self have been influenced by the internet and digitality. I introduce pre-existing work to argue that distancing the self from technological means can risk "shifting out" [16] our perspectives and as a result, dissociating from attacks to selfhood online. I discuss existing work on informational privacy and consent and ask if it is an adequate concept to cover such privacy attacks involving the body. Linking previous feminist theory on bodily integrity, I put forward the case for a related concept, *online bodily integrity*, which could provide language for non-consensual uses of the body likeness online. I draw on the previous argument that a distanced frame of reference for such crimes will affect our empathy and consideration for victims and we should consider our online identity as part of a larger idea of self. I conclude that when we distance ourselves from our digital counterparts, we inadvertently surrender our bodies' ownership to the internet.

2 Methods and Terminology

I will be following the concept of the embodied self, in which our lived experiences and our physicality is considered a part of the self, as opposed to a mind-body dichotomy. Further, I will be drawing upon a postmodern and symbolic interactionalist[1] influenced view of the construction of the self, namely that the "self is conceptualised as something presented to the community at large and thus accessible through interaction" [27] and further that this self is "shifting and unanchored" [31]. Put simply, this will mean a self that is highly shaped by social interaction and not a fixed entity. The online self, subsequently, I will define as the version of the self presented during any kind of social interaction online.

3 Deepfakes

The word deepfake originates from 'deep learning' and 'fake', where deep learning is a sector of machine learning which uses artificial neural networks. This can be supervised, which uses a labelled dataset to train on and reinforce the learning based on known results, or unsupervised which trains on an unlabelled dataset. Deepfakes are notable due to their use of generative adversarial networks (GANs). GANs work by training two neural networks together, a 'generator' and a 'discriminator', where the generator replicates input images and tries to 'fool' a generator, whose goal is to distinguish between the original input and the generator's replication. This uses a combination of supervised and unsupervised techniques [33]. Visual effects software, such as Photoshop, predates deepfake technology and has been used to create non-consensual imagery, but never at this scale. The developments in deep learning have caused a shift towards video content in particular that reaches a new level of sophistication and believability. Additionally, this deepfake technology is easier to access, often free and does not require the level of skill or time that previous image manipulation might require [22].

3.1 Deepfake Pornography

As referenced in the introduction, deepfake pornography forms the majority of the use of this technology. Within this exists the use of deepfakes for revenge porn, which is the practice of circulating sexually explicit imagery with the intention to humiliate, threaten or cause reputational damage [20]. The practice of creating such non-consensual deepfakes is believed to have begun with FakeApp, which was shared and used to create deepfake pornography of celebrities on Reddit forums in 2017 [10]. In turn, this software was then used to create doctored images of individuals personally known to perpetrators. This was originally uncovered by reporter Samantha Cole, whose work led to Reddit banning the practice on the platform [7]. Delfino highlights the sheer mass of online imagery at the disposal of perpetrators to use [10]. Whether 'innocent' pictures from a Facebook account or existing sexually explicit images - those who desire to use

---

[1] Symbolic interactionalism is a "theoretical perspective in sociology that addresses the manner in which society is created and maintained through face-to-face, repeated, meaningful interactions among individuals" [6].

this technology to violate victims' privacy do not have to put significant effort into sourcing the little material needed for deepfake models to replicate.

3.2 Existing Literature on Deepfakes

The threat to privacy from deepfakes has been discussed extensively since their inception. Alongside pornography, the threat to security by politically motivated deepfake videos has been highlighted repeatedly in the ensuing literature since 2017 [22].

In the *Distinct Wrong of Deepfakes*, de Ruiter argues that deepfakes are "morally suspect, but not inherently morally wrong" [9]. De Ruiter focuses on deontological reasoning to reach her conclusion, concentrating on the ability of deepfake imagery to deceive. She concludes that the existence of 'morally positive' uses for deepfake imagery, such as image distortion to protect the identity of persecuted members of the LGBT+ community in Chechnya, that do not violate moral norms subsequently indicate that it is not fundamentally a morally wrong technology. This reasoning focuses strongly on deception as the main vice, or moral wrong, stemming from deepfakes. However, I would urge that alongside deception, the vices of humiliation, intimidation and privacy and trust violations are strongly associated with the use of deepfakes. Further, I argue that focusing on the degree of believability in deepfake pornography is a well-intended but harmful distraction from the actual effect it causes on the victims. As exhibited by Farokhmanesh, and by de Ruiter when citing Cook, victims and assailants alike know that the end purpose of using deepfake imagery is not to trick recipients into believing it is genuine imagery of the victim [12] [8]. Repeated focus on deepfake deception and the convincingness of the fakery in this instance inadvertently leans into discussing the intricacies of the technology itself and shifts focus away from the results for the victims, themselves and their bodies. Victims' accounts of deepfake revenge porn suggest feelings of humiliation and violation of privacy as the primary intention rather than deception. Blaire, a targeted Twitch[2] streamer in Farokhmanesh's piece, remarks "Even though 'it's not my body' it might as well be. [It was] the same feeling-a violation that comes with seeing a body that's not yours being represented as yours" [12]. Equally as harrowing is the case of Rana Ayyub, a noted Indian journalist who was attacked with a deepfake pornography campaign as retribution for speaking in the defence of an 8-year-old victim of rape in a highly publicised and controversial crime in India. Ayyub described the physical distress she faced when viewing the deepfake version of herself online as so severe she was later hospitalised [3].

As is often the case with technological advances which have created new issues rapidly, it has been hard for legislators worldwide to respond quickly to the threat of this specific act. Revenge porn itself does have specific legislation in numerous countries, but the effectiveness of this legislation is debated and contended by victims [13]. In Japan for example, Matsui argues that haste in passing revenge porn legislation resulted in an act that is overly broad, lacks

---

[2] Twitch is an online video game live streaming platform.

sufficient punishment and does not provide much further protection than existing laws against "obscene" content in Japan [19]. Amongst all legislation, a common problem persists. The internet's lack of borders means victims often have to pursue justice in a different jurisdiction or legal system, and even if they succeed in their pursuit, removing content from the internet is a "Sisyphusian"[3] [12] task. Deepfake revenge porn content specifically has even less legal protection. Mania found that amongst EU countries, most treated deepfake revenge porn as "not a sexual offence, but only a minor privacy violation offence" [18].

As is evidenced, the aftermath of these attacks for victims is often life-changing and amplified by the lack of justice and paths available to seek legal retribution or erasure of the material online. Victims often have to live with the knowledge that their body likeness is still being circulated on the internet. To categorise this as an attack on someone's data, a 'minor' privacy offence, as Mania observed, does not seem sufficient to articulate the moral wrong faced.

In the next two sections, I propose that there are two notable factors that could contribute to the way deepfake revenge porn is received, a difference in the sense of self on the internet which causes these crimes to be under-acknowledged, and informational privacy's failure to intersect with crimes to online bodies. As discussed later, the prevalence of women being the primary targets of these attacks is also a factor in their perception, hence feminist theory has much to lend to considering this act.

Drawing on the victim's statements, there is a commonality in how they emphasise how an online attack feels to their actual self and body. This is evidenced in victims' accounts of revenge porn online without deepfakes, as Patella-Rey observes in actress Jennifer Lawrence's account of revenge porn "she does not distinguish between the representations of her body and the body itself." [25]. In the next section, I contextualise the relationship between the self, the body and the internet and highlight how this can create a disparity between the victims' experience and the perception of the outside world.

4 The Self and the Internet

Many theorists, particularly across sociology and philosophy, have argued the integration of the internet into our daily lives has warped our perception of self. Most relevant to this paper is the debate over the representation of 'the self' online being real and true, or a more hollow façade. As Zhao notes, there is a wider preoccupation with the presentation of the self online rather than the conception of the self in online spaces [34]. Zhao's version of Cooley's 20th century 'looking glass' concept, which notably was published in 2005 before the widespread use of social media today, forwards a version of the online self built on symbolic interactionist thinking, which he further links to Foucault and other postmodernist ideals on the self as a product of social interaction:

---

[3] A laborious and unfruitful process. Refers to the Greek mythological figure of Sisyphus, who was doomed to roll a boulder up a hill every day, for it only to roll back down again as it approached the top.

> *As a product of pure linguistic manipulations, the digital self illustrates the power of language and discourse, and the ease with which individuals change their online identities attests to the fluidity and inconstancy of the self.* [34]

Zhao argues here that the volatility of the online landscape means more flux in the self on the internet. A repeated argument for this is our ability to erase versions of ourselves online. This might have seemed more true in 2005 than today. It has become increasingly impossible to truly erase a digital footprint, whether it was voluntarily uploaded or non-consensually, even with the newer 'Right to Forget' legislation in the EU [2]. Thus, with increased digitalisation, the representation and conception of ourselves online have become more multifaceted and vast, and more interlaced with the self. Englezos argues the digital self is often considered "a third dimension to the individual: We are now mind, body and digitality" [11]. She further argues that the distinct separation of digitality from the self "undermines the real person's autonomy by allocating rights or opportunities according to digital proxies rather than the real person" [11]. Englezos further invokes Bruno Latour's "shifting out" [16], she argues that continual shifting of frames of reference for internet selfhood contributes towards this practice of disassociation.

If we internalise the postmodern view of self, the dissociation of this further online facet of oneself is consistent with an ever-shifting self. Zhao's "looking glass" and Englezos's arguments towards dissociation both could be said to explain a different perception when this online self is attacked or manipulated without consent. Referring back to deepfake pornography, Meskys et al. observe "a great number of the online community is rather indifferent" to the act [20]. Blaire, the targeted Twitch streamer, reports a similar level of ambivalence from her followers who questioned her distress when the content was clearly not her, which Farokhmanesh describes as a "bad faith" interpretation deliberately missing the "real harms" [12]. I argue that the distance between the online and the physical self is exhibited in the reaction to such crimes. This is consistent with a shifting or fluid self, where the context of the social interaction is dictated by the surrounding norms. In the case of deepfake pornography, the comparative ease of access to the tooling and surplus of similar uses sets the expectation for the perpetrator. Dissociation from the selfhood of the victim allows perpetrators to distance themselves from consequences and further, inflict real harm with little effort.

Comparing these sentiments to the victims' statements shared in the previous section, it is arguable that this ambivalence or lack of relation to the self is not felt by the victims, who face a significant amount of emotional trauma. Their sense of privacy is not concentrated purely on this sole instance of their online data and imagery, it bleeds into their everyday life and larger sense of self. Thus, there is a clear suggestion that there needs to be further thought into the commonality between an online privacy violation and a physical privacy violation. In the next section, I evaluate this idea by considering the relationship between self, body and data in the context of informational privacy.

5 Informational Privacy

Informational privacy is the most commonly linked form of autonomy discussed in relation to the internet. A simple definition is "the ability to determine what others do with your information" [32]. Critique of informational privacy, particularly evaluation within feminist thinking, has proposed that informational privacy fails to incorporate existing ideas around consent and views one's data online as a depersonalised entity. Referring back to the previous section, this depersonalisation could be said to be aligned with the sense of dissociation one can experience with their online self. Relevant to the scope of this paper, it must be questioned whether non-consensual replications of the body online are best categorised under *information* and if not, subsequently whether informational privacy is adequate to cover this kind of violation.

In *Privacy in Public*, Nissenbaum highlights the longstanding belief in philosophical and legal theory that publicly available information does not warrant the same privacy norms as private information [23]. Considering the online self, there are nuances in the degree to which our information is truly publicly available, available to a wider group of connections or 'private' between the individual and the online platform. As Nissenbaum acknowledges heavily, 'private' information is somewhat a fallacy given the ever-growing chain of access to such data, including cloud storage and third parties. She argues that advances in information technology facilitate vastly larger practices of surveillance and thus we need to reevaluate this stance. The distinction between the public and private divide can be felt acutely in instances of deepfake revenge porn, where benign public images can be warped into pornography, and where sex in a majority of societies worldwide is considered a highly private act.

Nissenbaum has also provided extensive work discussing the nuances of informed consent and introduced the concept of contextualised consent online [24]. As Bernal paraphrases: "a piece of information that in one moment appears innocuous can in another be of the greatest sensitivity and importance" [4]. The efficiency of such consent online is a much-discussed debate. Some argue that a more nuanced version of informed consent is needed, whilst others argue that consent is akin to a rouse to disguise the ultimate motivations of those who seek to use our data. When discussing consent in the context of egregious privacy violations, for which I would categorise deepfake revenge porn of any kind, I would argue that informed consent renders idle. As evidenced in non-consensual deepfakes, the consent to a publicly available picture or video is weaponised into a violation almost nobody would consent to in any form.

The uncomfortable categorisation of personal imagery violations under informational privacy could be said to be a problem with our perception and categorisation of what 'information' means. In a more recent analysis of Nissenbaum, Mai discusses the symbiotic nature of information. Informational privacy, he argues, has largely not contended with what personal information actually consists of, "leaving an understanding of personal information in which information *just is*" [17]. He further argues that the bigger concern should be with the situation in which the meaning of the data is created, thus the importance resides in the relationship between the two and not the information itself. He furthers Nissenbaum's stance that context is the key component of meaningful informational privacy. Mai's analysis provides a foundation for which we can consider the gravity of the relationship as a driving motivation in treating a privacy violation. In the situation of non-consensual deepfake porn, there is a clear, acute imbalance in the social interaction.

5.1 Informational Privacy and the Body

Alongside the importance of social interaction in thinking about informational privacy, criticism has also highlighted the difference in privacy expectations when we begin to consider embodied approaches to data. Several authors have argued that informational privacy is inflexible when considering it in the context of bodies online. In *The body as data in the age of information*, van der Ploeg reflects on the gulf between how we protect bodies from intrusion and how we view data protection:

> *In the case of the body-as-information, the problem is that we have very different regimes for protecting bodies and for protecting information from unjustified access and intrusion, however "personal" that information may be. Whereas in the first case, the very integrity of the body and issues of self-determination are at stake, in the second, the far weaker concepts of informational privacy and personal data protection apply.* [30]

She concludes that in the case of intrusion online concerning personal data related to the body, informational privacy becomes a weaker concept to apply. She considers in depth how the body and the self have become coded as information, concluding that we should take a "subtle but highly consequential" [30] shift in the way we view our materiality. With a different lens, Peña and Varon intertwine the long, existing dialogue on consent in feminist work, particularly queer feminist theory, with information privacy in *Consent to our Data Bodies*. They argue there are significant learnings that can be applied to data consent, and introduce the idea of a "data body" [26], highlighting the historical feminist relation of bodily integrity and consent to online versions of consent. On Nissenbaum's idea that the internet is a uniquely considered realm for which we "define by the technological infrastructures and protocols of the Net, for which a single set of privacy rules can, or ought to, be crafted" [26] Peña and Varon argue that Nissenbaum's separation is misjudged given it "seems obvious that our offline and online life is radically interconnected" [26]. They further argue that this can be seen in Elinor Carmi's work, where she argues that data is wrongly perceived to be a fixed piece of personal property, which is contrary to the reality of the fluidity of this relationship between data and self [5]. This complements Zhao's analysis of the fluid state of the online self and furthers that perceiving data with rigidity is in conflict with the actuality of how we experience our bodies online.

This section has covered varied reflections on the application of informational privacy to the self on the internet and has surfaced areas of ambiguity on the relationship between data and body which could benefit from further thought or a different conceptual analysis. The concept of "data bodies" in particular introduces an interesting parallel between bodily integrity and the online self in the context of deepfake pornography.

6 Online Bodily Integrity

Bodily integrity is concerned with the right to maintain autonomy with regard to intervention or influence over one's body. This is a key factor in medicine when considering patients' rights to control medical intervention and choose their desired treatment [14]. It is also more commonly linked to feminist theory, thoroughly interwoven in the history of patriarchal intervention in women's bodies. Restricting women's right to abortion is a resonant example of conflict over bodily integrity and autonomy. Kovacs relates this concept back to the victims of image abuse online:

> *To name but one example, victims of the non-consensual sharing of sexual images generally do not describe the harms they experience in terms of data protection or even privacy violation. Rather, they describe the harms as similar to those arising from sexual assault: as a violation of bodily integrity.* [15]

The relation supposed here between bodily integrity and non-consensual sexual imagery is echoed in others' work. Patella-Rey has discussed the concept of bodily integrity specific to non-consensual pornography, arguing it provides a more fitting concept to the experience of victims. His analysis is focused on the inadequacies of existing legal privacy theory with regard to women's bodies and is deeply applicable here when considering informational privacy and the body. He also highlights the importance of shifting perceptions of the self online, drawing on Allucquère Rosanne Stone's argument that digital interactions "are forcing us to reimagine the boundaries of where the self begins and ends" [25]. Subsequently, he argues we should consider techno-feminist perspectives when considering such violations, which "seek to prevent the technologically mediated facets of the self from becoming sites of domination, paying particular attention to where technology intersects with bodies (as is the case with nude/sexual image sharing)" [25].

Another related work on bodily integrity on the internet is "bodyright", a project by the United Nations Population Fund, which invites participants to bodyright their images online by displaying a symbol on pictures uploaded online [1]. The project argues that the misogyny and violence online displayed in image abuse such as deepfake pornography is of less importance to online platforms than corporate copyrights and such, we should commodify our bodies in similar ways in order to be taken as seriously. One can not fully understand the phenomenon of deepfake revenge porn without taking into account that it is a vastly gendered issue. The long history of the sexualisation of women's bodies is a constituting factor of this practice in the modern day. Relating the feminist view of bodily integrity to the crime of deepfake revenge porn, there is a clear overlap between the perception of victims' rights to autonomy and privacy. However, it cannot be ignored that there is a distinguishable difference between a violation of a physical body and online abuse of the body's likeness. Whereas many would use this difference to minimize the effect of revenge porn, I instead argue that we should consider the nuance of both and unite these concepts into an idea of *online bodily integrity*. Extending Patella-Rey's proposition of bodily integrity as a framework, *online bodily integrity* can concern 'one's right to exist online without non-consensual interference with representations of their body'. We can consider this the online application of the right to self-ownership and self-determination of our body, which extends to protection against harassment and deception. Considering this as a right more tightly linked to our physical self and not purely a one-dimensional representation of our data or information encompasses the gravity of the violation and could provide greater understanding to those who wish to combat such crimes, through legislation or other social action.

7 Conclusion

If we consider an aspect of autonomy to be the ability to be able to carry out one's life without manipulation, and further if we accept the feminist philosophical stance to incorporate 'social freedom' into autonomy, the classification of deepfake revenge porn as a gross violation of privacy is hard to refute [4]. Dissociating the mind and physical body away from our digital selves online allows us to frame privacy attacks such as deepfake revenge pornography as a crime only against a distanced, online self and diminish the real aftermath that victims face in reality. Thus, I have argued that *online bodily integrity* is a more fitting concept to categorise the type of privacy threatened here. By considering the type of privacy violation through a more modern prism, the difficulties of fitting this act into the category of a data or information breach could be avoided and more victims could feel properly heard. It is of crucial importance that we put the experiences of victims at the forefront of social and legislative action on this matter and let them retain a sense of ownership of themselves and their bodies online.

References

1. A new copyright for the human body. Bodyright - own your body online: bodily integrity: UNFPA. https://www.unfpa.org/bodyright
2. Art. 17 GDPR right to erasure ('right to be forgotten') (2023). https://gdpr.eu/article-17-right-to-be-forgotten/
3. Ayyub, R.: I was the victim of a deepfake porn plot intended to silence me. Huffington Post (2018). https://www.huffingtonpost.co.uk/entry/deepfake-porn_uk_5bf2c126e4b0f32bd58ba316

4. Bernal, P.: Privacy, autonomy and the internet, pp. 24–52. Cambridge Intellectual Property and Information Law, Cambridge University Press (2014). https://doi.org/10.1017/CBO9781107337428.003
5. Carmi, E.: Do you agree?: What #metoo can teach us about digital consent. Open Democracy (2018). https://www.opendemocracy.net/en/digitaliberties/what-metoo-can-teach-us-about-digital-consent/
6. Carter, M.J., Fuller, C.: Symbols, meaning, and action: the past, present, and future of symbolic interactionism. Curr. Sociol. **64**(6), 931–961 (2016). https://doi.org/10.1177/0011392116638396
7. Cole, S.: Ai-assisted fake porn is here and we're all fucked. Motherboard (2017). https://www.vice.com/en/article/gydydm/gal-gadot-fake-ai-porn
8. Cook, J.: Here's what it's like to see yourself in a deepfake porn video (2019). https://www.huffingtonpost.co.uk/entry/deepfake-porn-heres-what-its-like-to-see-yourself_n_5d0d0faee4b0a3941861fced
9. de Ruiter, A.: The distinct wrong of deepfakes. Philos. Technol. **34**(4), 1311–1332 (2021). https://doi.org/10.1007/s13347-021-00459-2
10. Delfino, R.A.: Pornographic deepfakes: the case for federal criminalization of revenge porn's next tragic act. Fordham Law Rev. **88**, 887 (2019). https://doi.org/10.2139/ssrn.3341593
11. Englezos, E.: Sign of the times: legal persons, digitality and the impact on personal autonomy. Int. J. Semiot. Law - Rev. int. de Sémiotique juridique 1–16 (2022). https://doi.org/10.1007/s11196-022-09925-2
12. Farokhmanesh, M.: The debate on deepfake porn misses the point. Wired (2023). https://www.wired.com/story/deepfakes-twitch-streamers-qtcinderella-atrioc-pokimane/
13. Hao, K.: Deepfake porn is ruining women's lives. now the law may finally ban it. MIT Technol. Rev. **12** (2021). https://www.technologyreview.com/2021/02/12/1018222/deepfake-revenge-porn-coming-ban/
14. Herring, J., Wall, J.: The nature and significance of the right to bodily integrity. Cambridge Law J. **76**(3), 566–588 (2017). https://doi.org/10.1017/S0008197317000605
15. Kovacs, A.: Towards an embodied approach to data. Internet Democracy Project. https://internetdemocracy.in/bodies-and-data
16. Latour, B.: Pandora's Hope: Essays on the Realities of Science Studies. Harvard University Press, Cambridge, Massachusetts (1998). https://doi.org/10.5860/choice.37-1511
17. Mai, J.E.: chap. Situating Personal Information: Privacy in the Algorithmic Age. The MIT Press (11 2019). https://doi.org/10.7551/mitpress/11304.001.0001
18. Mania, K.: Legal protection of revenge and deepfake porn victims in the European union: findings from a comparative legal study. Trauma, Violence, & Abuse (2022). https://doi.org/10.1177/15248380221143772
19. Matsui, S.: The criminalization of revenge porn in japan. Wash. Int. Law J. **24**(2), 289–317 (2015)
20. Meskys, E., Liaudanskas, A., Kalpokiene, J., Jurcys, P.: Regulating deep fakes: legal and ethical considerations. J. Intellect. Property Law Pract. **15**(1), 24–31 (2020). https://doi.org/10.1093/jiplp/jpz167
21. Newen, A.: The embodied self, the pattern theory of self, and the predictive mind. Frontiers Psychol. **9**, 2270 (2018). https://doi.org/10.3389/fpsyg.2018.02270
22. Nguyen, T.T., et al.: Deep learning for deepfakes creation and detection: a survey. Comput. Vis. Image Underst. **223**, 103525 (2022). https://doi.org/10.1016/j.cviu.2022.103525

23. Nissenbaum, H.: Protecting privacy in an information age: the problem of privacy in public. Law Philos. **17**(5–6), 559–596 (1998). https://doi.org/10.2307/3505189
24. Nissenbaum, H.: Privacy in Context. Stanford University Press, Redwood City (2009). https://doi.org/10.1515/9780804772891
25. Patella-Rey, P.: Beyond privacy: bodily integrity as an alternative framework for understanding non-consensual pornography. Inf. Commun. Soc. **21**(5), 786–791 (2018). https://doi.org/10.1080/1369118X.2018.1428653
26. Peña, P., Varon, J.: Consent to our data bodies: Lessons from feminist theories to enforce data protection. Coding Rights (2019). https://codingrights.org/docs/ConsentToOurDataBodies.pdf
27. Riggs, A., Turner, B.: The sociology of the postmodern self: intimacy, identity and emotions in adult life. Aust. J. Ageing **16**(4), 229–232 (1997). https://doi.org/10.1111/j.1741-6612.1997.tb01061.x
28. Silberling, A.: Kickstarter shut down the campaign for AI porn group unstable diffusion amid changing guidelines. TechCrunch (2022). https://techcrunch.com/2022/12/21/kickstarter-shut-down-the-campaign-for-ai-porn-group-unstable-diffusion-amid-changing-guidelines
29. Sontag, S.: On Photography. Farrar, Straus & Giroux (1977)
30. van der Ploeg, I.: Routledge Handbook of Surveillance Studies, chap. The body as data in the age of information. Routledge (2012). https://doi.org/10.4324/9780203814949
31. Walker, K.: "it's difficult to hide it": the presentation of self on internet home pages. Qual. Sociol. **23**, 99–120 (2000). https://doi.org/10.1023/A:1005407717409
32. Warner, R., Sloan, R.: Self, privacy, and power: Is it all over? Tul. J. Technol. Intell. Prop. **17**, 61 (2014). https://doi.org/10.2139/ssrn.2401165
33. Westerlund, M.: The emergence of deepfake technology: a review. Technol. Innovation Manag. Rev. **9**, 40–53 (2019). https://doi.org/10.22215/timreview/1282
34. Zhao, S.: The digital self: through the looking glass of telecopresent others. Symb. Interact. **28**(3), 387–405 (2005). https://doi.org/10.1525/si.2005.28.3.387

The Promise and Peril of Responsible AI Principles

Retno Larasati[✉]

Knowledge Media Institute, The Open University, Milton Keynes, UK
`retno.larasati@open.ac.uk`

Abstract. This paper provides an examination of responsible AI principles by prominent tech companies, namely Microsoft, Amazon, and Meta. The company perspectives are discussed through an exploration of key AI principles and their alignment with the Partnership on AI (PAI) thematic pillars. By critically assessing the implementation of these principles, the paper underscores the need for further research to comprehensively evaluate their efficacy and impact in real-world contexts.

Keywords: AI Principles · Ethical AI · Responsible AI

1 Introduction

In recent years, the rapid advancements in Artificial Intelligence (AI) have brought about significant transformations in various domains of society. As AI technologies continue to proliferate, so does the importance of responsible AI deployment. With the rapid advancement of artificial intelligence (AI) technologies, there has been a growing concern about the ethical implications and societal impact of these powerful systems. In response to these concerns, various stakeholders, including companies, regulatory bodies, and advocacy groups, have emphasised the importance of ethical or responsible AI [5]. They have called for the development and implementation of principles and guidelines that ensure AI is used ethically, transparently, and with accountability.

The purpose of this paper is to examine the promise and peril associated with the implementation of responsible AI principles by major tech companies. By focusing on the commitments and actions of prominent companies in embracing responsible AI, we aim to evaluate the extent to which these principles translate into meaningful practices. This exploration will shed light on the challenges and gaps in implementing responsible AI, providing insights into the potential perils of relying solely on industry-led initiatives like the Partnership on AI (PAI).

In the subsequent sections, we will assess the implementation of AI principles, focusing on the potential perils and limitations. Through this exploration, we aim to contribute to the ongoing discourse surrounding the ethical deployment of AI and the necessity for robust accountability mechanisms. By critically examining the promises and perils of responsible AI principles, we hope to foster a greater understanding of the challenges and opportunities in realizing the full potential of AI technologies in an ethically sound manner.

2 Background

From the algorithms that recommend products to us on Amazon to the self-driving cars that are being developed by Google and other companies, AI is currently having a profound impact on the way we live and work. As AI becomes more powerful, it is important to consider the ethical implications of its use. AI systems can be used to discriminate against certain groups of people [14], invade our privacy [11], and even cause physical harm [8]. It is therefore essential that companies develop and implement responsible AI.

There are several definitions that organisations use to describe "responsible", "trustworthy", and "ethical AI". For example, the European Commission is using the term trustworthy AI, and defined trustworthy AI with three components that should be met throughout the system's entire life cycle: (1) it should be lawful, complying with all applicable laws and regulations (2) it should be ethical, ensuring adherence to ethical principles and values and (3) it should be robust, both from a technical and social perspective since, even with good intentions, AI systems can cause unintentional harm [7].

Microsoft and IBM, on the other hand, utilise the term Responsible AI for their AI principles and frameworks. They define Responsible AI as AI that is developed and used in a way that is ethical and beneficial to society. While their specific definitions may differ, the underlying principles remain consistent, emphasising the importance of ethical considerations, societal benefit, and accountability in the development and deployment of AI technologies.

The interchangeability of terms such as trustworthy, ethical, and responsible when describing a *'good'* AI reflects the multi-faceted nature of the principles and values underlying responsible AI development and deployment. Trustworthiness, ethics, and responsibility are interconnected aspects that collectively contribute to the development of AI systems that align with societal expectations and promote positive outcomes. We chose to focus on responsible AI because it encompasses a broader range of sub-topics, including trustworthiness and ethics, while also emphasizing the overall societal benefit.

3 Responsible AI Principles

There are different definitions of responsible AI, however, for this paper we will focuses on the following five key principles of responsible AI:

- **Fairness**: AI systems should not discriminate against individuals or groups of people.
- **Transparency**: AI systems should be designed and operated in a manner that enables users and stakeholders to understand the underlying mechanisms, processes, and decisions made by the AI.
- **Safety**: AI systems should be designed and tested to ensure that they are safe to use.
- **Privacy**: AI systems should not collect or use personal data in a way that is harmful to individuals or society.

– **Accountability**: There should be clear processes in place to hold companies accountable for the use of AI systems.

Companies play a pivotal role in shaping the AI landscape, and their commitment to responsible AI principles holds the promise of fostering ethical, transparent, and accountable AI technologies. To explore the AI principles of various companies, a systematic research methodology was employed. The following steps were undertaken to ensure a comprehensive analysis of companies' responsible AI practices:

Identification of Target Companies: A set of companies known for their involvement in AI research, development, or deployment was selected. The selection aimed to encompass a diverse range of sectors and geographical locations to capture a broad perspective on responsible AI principles. In this paper, we are focusing on the published principles and guidelines, as well as the practices, of MAMAA (Meta, Amazon, Microsoft, Alphabet, Apple) companies. The MAMAA companies, an updated version of FAANG (Facebook, Amazon, Apple, Netflix, and Google) that includes Microsoft instead of Netflix and reflects rebrandings of FAANG members Facebook and Google, stand at the forefront of AI development and deployment. We also added Deep Mind as a Google subsidiary, IBM, and OpenAI, as those companies are also at the forefront of AI development and deployment currently.

Preliminary Search: An initial search was conducted on prominent search engines, such as Google, to identify official company websites, press releases, blog posts, white papers, and other relevant sources that provide insights into the companies' AI principles.

Document Search Criteria: A set of keywords and search queries were developed based on the research objective. These search criteria were tailored to retrieve information related to the companies' AI principles and responsible AI guidelines. Examples of search queries included "Company X AI principles," "ethical AI Company Y," or "Company Z responsible AI guidelines."

Data Collection: Information was gathered from various sources, including official company statements, published reports, policy documents, and reputable news articles. Primary emphasis was placed on accessing official company resources to ensure accuracy and reliability.

Data Analysis: A thematic analysis approach was employed to identify common themes and key principles across the companies' AI principles. The analysis focused on categorising the principles according to the five key principles of responsible AI: fairness, transparency, security, privacy, and accountability.

Cross-referencing with Partnership on AI (PAI) Thematic Pillars: To provide another overview, the companies' AI principles were cross-referenced with the thematic pillars outlined by the Partnership on AI (PAI). The PAI is a non-profit coalition aimed at fostering responsible AI applications and promoting public comprehension of the field. In 2016, a collaborative effort was initiated by

Google, Facebook, Amazon, IBM, and Microsoft, to establish an AI partnership with the primary objective of promoting public comprehension of the field and formulating guidelines for future researchers to adhere to named [6]. Referred to as the Partnership on AI (PAI), or the Partnership on Artificial Intelligence to Benefit People and Society, PAI is a non-profit coalition that holds a steadfast commitment to fostering the responsible application of artificial intelligence. Apple then became a "founding" member of the board in 2017 [4].

Result In the Table 1 below we listed companies' AI principles and PAI thematic pillars, divided by the five key principles of responsible AI.

Even though Apple has published articles related to AI and machine learning in their "Machine Learning Journal" and "Apple Machine Learning Research," they do not have a specific standalone document or web-page dedicated to "AI principles" or "responsible AI guidelines".

Table 1. AI companies and their responsible AI principles

PAI and AI companies	Fairness, justice	Transparency, explainability	Security, safety	Privacy	Accountability
Partnership on AI	No bias, fairness, inclusivity	Transparency, explainability, interpretability, understandability	Safety, security, reliability, robustness	Privacy	Accountability
Meta (Facebook)	Fairness, inclusion	Transparency, control	Robustness, safety, security	Privacy	Accountability, Governance
Amazon	Fairness, no bias	Transparency, explainability	Security	Privacy	Governance
Microsoft	Fairness, inclusiveness	Transparency, intelligibility	Safety, reliability, security	Data privacy	Accountability
Apple	–	–	–	–	–
Google (Alphabet)	No unfair bias	Relevant explanations	Safety, security	Privacy	Accountability
Deep Mind (Google subsidiary)	No bias, no discrimination, fairness, inclusivity	Transparency		Privacy	Accountability
IBM	No bias, no discrimination, fairness	Transparency, explainability	Data security, robustness	Data privacy	Accountability
OpenAI	–	–	Safety	–	–

4 Discussion: The Promise and Peril of Responsible AI Principles

The MAMAA companies have a profound influence on AI technologies and their societal impact. Previous research based the partnership with the PAI as as an indication of a company's commitment to AI principles [10]. If we considered being a PAI partner as a signal of their commitment towards responsible AI, we can argued that seven out of eight companies (Meta, Amazon, Microsoft, Alphabet, Apple, Deep Mind, IBM), are publicly committed to their promise of responsible AI.

While the Partnership on AI (PAI) has been influential in encouraging responsible AI practices, it is important to acknowledge the flaws to use it as an indicator of true commitment to ethical AI. The issue of accountability arises when evaluating the implementation of the principles these companies published. The question of who accounts for the implementation of AI principles becomes crucial. For internal accountability mechanisms, companies can establish internal structures and mechanisms to ensure accountability in responsible AI principle implementation. These may include dedicated teams, governance boards, and compliance frameworks responsible for monitoring and enforcing adherence to responsible AI practices within the organisation [13].

However, according to Benkler, the influence of industry in shaping AI regulations, or in this case also *self-regulation*, can lead to a self-serving approach that prioritises corporate interests over broader societal concerns. Industry-led initiatives may be driven by a desire to avoid stringent regulations that could impact their business models or limit their competitive advantages. This can result in the development of AI guidelines and principles that are narrow in scope, favor industry interests, and neglect critical ethical considerations [1].

Critics from internal side came in 2020. Access Now, an international organisation advocating for digital rights and human rights, made the decision to resign from the PAI as a form of protest. This decision was motivated by their perception that the businesses affiliated with the coalition had not made significant progress in addressing the ethical challenges of AI. Access Now also expressed concerns about the PAI's failure to incorporate viewpoints advocated by civil society organisations, which are critical in shaping comprehensive and inclusive AI policies [15]. It is important to incorporate independent and critical perspectives in shaping AI policies. It underscores the need for academic researchers, public interest organisations, and civil society to maintain autonomy and avoid undue industry influence. By doing so, the development of AI governance frameworks can be more inclusive, transparent, and responsive to the broader societal interests [12].

External entities, including regulatory bodies, industry associations, and civil society organisations, play a significant role in holding tech giants accountable for their AI principle implementation. Regulatory bodies, such as government, in particular, can establish guidelines and frameworks that require companies to adhere to responsible AI practices, conduct audits, and impose penalties for non-compliance. Industry associations and civil society organisations can advocate for ethical AI practices and scrutinise companies' adherence to their stated commitments [3, 9]

In the recent years, there has been a growing emphasis on the responsible and ethical use of AI, with various countries and regions introducing or strengthening AI-related regulations and guidelines, such as, GDPR in 2018 and AI Act in 2021 for European Union Countries. Government policies promoting transparency, fairness, and accountability in AI technologies may have influenced companies' decision to publicly articulate their AI principles. By aligning their practices with government policies and public expectations, companies can demonstrate their commitment to responsible AI deployment.

The release of AI principles by companies like Meta (formerly known as Facebook) in 2021 and other tech giants may be attributed, at least in part, to the growing public backlash and increased scrutiny surrounding the ethical implications of AI technologies. The Cambridge Analytica scandal and subsequent inquiries into data privacy violations brought significant attention to the responsible use of AI within the company [11]. It is plausible that the growing public demand for transparency and accountability in AI systems played a role in Meta's decision to release their AI principles at this juncture. Similarly, Amazon's release of their AI principles at the end of 2022 can be seen as a response to the evolving landscape of public expectations and government policies. Amazon's infamous biased hiring system became a spotlight for AI fairness conversation for several years [2].

The formation of the PAI included participation from major tech giants, and solely relying on industry-led initiatives like the Partnership on AI (PAI) as indicators of responsible AI commitment is not ideal. It is indeed noteworthy that several companies, including Google, Microsoft, IBM, and others, released their AI principles around the same time as the initiation of the Partnership on AI (PAI). This simultaneous release of AI principles may indicate that the formation of the PAI, as a collaborative effort between major industry players, prompted a collective response to address the ethical challenges associated with AI.

However, recent developments have raised questions about the enduring nature of these commitments. For instance, both Meta and Microsoft made headlines for laying off their ethics and responsible technology teams as part of their massive layoffs affecting thousands of employees[1]. These actions raise concerns about the long-term sustainability and genuineness of their commitment to responsible AI. The decision to downsize or eliminate dedicated teams focused on ethical AI may signal a shift in priorities or a potential weakening of their dedication to addressing the ethical implications of AI technologies.

5 Conclusion: The Future of Responsible AI

Companies have a responsibility to develop and use AI responsibly. This means that companies should not only adopt a set of AI principles, they should commit to uphold. Companies should conduct regular ethics audits to assess the ethical implications of their AI systems. These audits should be conducted by independent experts and released to the public or reported to the government bodies. Inside the companies, they should create a culture of ethical AI within

---

[1] https://www.washingtonpost.com/technology/2023/03/30/tech-companies-cut-ai-ethics/.

their organisation. This means that all employees should be aware of the ethical principles of AI and how to apply them in their work.

Not only AI companies, government AI policies also play a vital role in shaping the responsible development and deployment of AI technologies. Governments have to recognised the need to strike a balance between fostering innovation and safeguarding societal interests. By establishing clear guidelines, regulations, and strategies, governments can foster a collaborative ecosystem that encourages innovation, while ensuring that AI technologies are developed and used in a responsible, fair, and accountable manner, ultimately benefiting society as a whole.

References

1. Benkler, Y.: Don't let industry write the rules for AI. Nature **569**(7754), 161–162 (2019)
2. Dastin, J.: Amazon scraps secret AI recruiting tool that showed bias against women. In: Ethics of Data and Analytics, pp. 296–299. Auerbach Publications (2022)
3. Floridi, L., et al.: An ethical framework for a good AI society: opportunities, risks, principles, and recommendations. In: Ethics, governance, and policies in artificial intelligence, pp. 19–39 (2021)
4. Forbes: Why apple joined rivals amazon, google, microsoft in ai partnership. https://www.forbes.com/sites/aarontilley/2017/01/27/why-apple-joined-rivals-amazon-google-microsoft-in-ai-partnership/ (2017), (Accessed on 06/13/2023)
5. Greene, D., Hoffmann, A.L., Stark, L.: Better, nicer, clearer, fairer: a critical assessment of the movement for ethical artificial intelligence and machine learning (2019)
6. Guardian, T.: 'partnership on AI' formed by google, Facebook, Amazon, Ibm and Microsoft — artificial intelligence (AI) — the guardian. https://www.theguardian.com/technology/2016/sep/28/google-facebook-amazon-ibm-microsoft-partnership-on-ai-tech-firms (2016). Accessed 13 June 2023
7. HLEG, E.: Ethics guidelines for trustworthy AI — shaping europe's digital future. https://digital-strategy.ec.europa.eu/en/library/ethics-guidelines-trustworthy-ai (2019). Accessed 13 June 2023
8. Jenssen, G.D., Moen, T., Johnsen, S.O.: Accidents with automated vehicles-do self-driving cars need a better sense of self? In: Proceedings of the 26th ITS World Congress, Singapore, pp. 21–25 (2019)
9. Jobin, A., Ienca, M., Vayena, E.: The global landscape of AI ethics guidelines. Nat. Mach. Intell. **1**(9), 389–399 (2019)
10. de Laat, P.B.: Companies committed to responsible AI: from principles towards implementation and regulation? Philos. Technol. **34**, 1135–1193 (2021)
11. Lauer, D.: Facebook's ethical failures are not accidental; they are part of the business model. AI Ethics **1**(4), 395–403 (2021)
12. Ochigame, R.: The invention of 'ethical AI': how big tech manipulates academia to avoid regulation. Economies virtue **49** (2019)
13. Scherer, M.U.: Regulating artificial intelligence systems: risks, challenges, competencies, and strategies. Harv. JL Tech. **29**, 353 (2015)

14. Stahl, B.C., Schroeder, D., Rodrigues, R.: Unfair and illegal discrimination. In: Ethics of Artificial Intelligence: Case Studies and Options for Addressing Ethical Challenges, pp. 9–23. Springer (2022). https://doi.org/10.1007/978-3-031-17040-9_2
15. VentureBeat: Access now resigns from partnership on AI due to lack of change among tech companies — venturebeat. https://venturebeat.com/ai/access-now-resigns-from-partnership-on-ai-due-to-lack-of-change-among-tech-companies/ (2020). Accessed on 13 June 2023

Accelerating Machine Learning Primitives on Commodity Hardware

Roman Snytsar[✉][iD]

AI and Research, Microsoft, Redmond, WA 98052, USA
Roman.Snytsar@microsoft.com

Abstract. Sliding Window Sum algorithms have been successfully used for training and inference of Deep Neural Networks. We have shown before how both pooling and convolution 1-D primitives could be expressed as sliding sums and evaluated by the compute kernels with a shared structure.

In this paper, we present an extensive study of the Sliding Window convolution technique as a more efficient alternative to the commonly used General Matrix Multiplication (GEMM) based convolution in Deep Neural Networks (DNNs). The Sliding Window technique addresses the memory bloating problem and demonstrates a significant speedup in 2-D convolution. We explore the performance of this technique on a range of implementations, including custom kernels for specific filter sizes.

Our results suggest that the Sliding Window computation kernels can outperform GEMM-based convolution on a CPU and even on dedicated hardware accelerators. This could promote a wider adoption of AI on low-power and low-memory devices without the need for specialized hardware. We also discuss the compatibility of model compression methods and optimized network architectures with the Sliding Window technique, encouraging further research in these areas.

1 Introduction

In recent years, there has been significant progress in machine learning (ML) research, with breakthroughs in deep learning, natural language processing, and computer vision. A Deep Neural Network (DNN) is one of the most significant tools of a ML scholar [17]. DNNs are constructed from multiple layers that transform the data sequentially via operations such as pooling, convolution, and activation. In most successful DNNs, the greater portion of computational resources is consumed by performing convolution.

A popular implementation of convolutional layers is expanding the input into a column matrix form (im2col) and then calling a highly tuned General Matrix Multiplication (GEMM) procedure from the existing linear algebra library such as BLIS [26] or MKL [28]. Since the hardware optimized GEMM implementations exist for every standard CPU, graphics processing unit (GPU), or digital signal processor (DSP), the im2col approach has been highly successful in DNN frameworks such as Caffe [15], Torch [5] and ONNX [3].

However, these advances have primarily benefited large corporations, and research institutions with access to massive computational resources. The democratization of AI on low power and edge devices aims to bring the benefits of AI to a wider audience, including small businesses, individual users, and the billions of Internet of Things (IoT) devices. Edge devices, such as smartphones, wearables, and IoT sensors, are often resource-constrained, with limited processing power, memory, and battery life.

One major challenge in deploying AI on edge devices is the size of deep learning models, which can be hundreds of megabytes or even gigabytes. The im2col conversion further increases the memory footprint of the input matrix and reduces data locality. For a convolution with a filter size k, the column matrix is k times larger than the original input tensor. A lot of research effort has been put into applying the GEMM routines to the smaller intermediate data structures [1,27] or even to the original input data [29].

To reduce the memory requirements on edge devices and improve performance, researchers have been exploring various techniques, including model compression [4,6–10,12,18,30,33], network optimization [11,13,22,25,32], and hardware acceleration [2,14,16,19,20,31].

2 Experiments

Earlier we proposed a new algorithm for performing convolution. The Sliding Window technique [24] replaces GEMM with a novel computation kernel that operates on the unmodified input and eradicates the memory bloating problem. The speedup of 1-D convolution we have observed when compared to the baseline *MlasConv* procedure was roughly proportional to the logarithm on the filter width [23].

In this paper we present the extension of the Sliding Window algorithms to the more practical 2-D cases. There are three different implementations on the 2-D sliding convolution. The kernel sizes up to 17 are handled by the straightforward version of the Vector Slide algorithm. Kernels of larger width do not fit into the hardware vector and require a special version that operates on multiple hardware vectors treating them as a single long compound vector. Both generic versions perform redundant shuffles, so for filter widths 3 and 5 we implemented custom kernels with optimal number of operations Fig. 1.

We have run experiments on an Azure node with 16 cores of Intel(R) Xeon(R) Platinum 8272CL CPU and 32 GB of RAM. The speedup is measured compared to the ONNX *MlasConv* calls. All tests have been run in a single-core configuration to exclude the effects of the task scheduling delays.

The 2-D Sliding Window convolution exhibits the same roughly logarithmic speedup in correlation to the filter size. The zigzag pattern at the larger filter sizes is related to the alignment of the compound vector to the hardware vector length.

Custom implementations are indeed faster than their generic counterparts. Generating custom kernels at run time might improve the performance for every filter size.

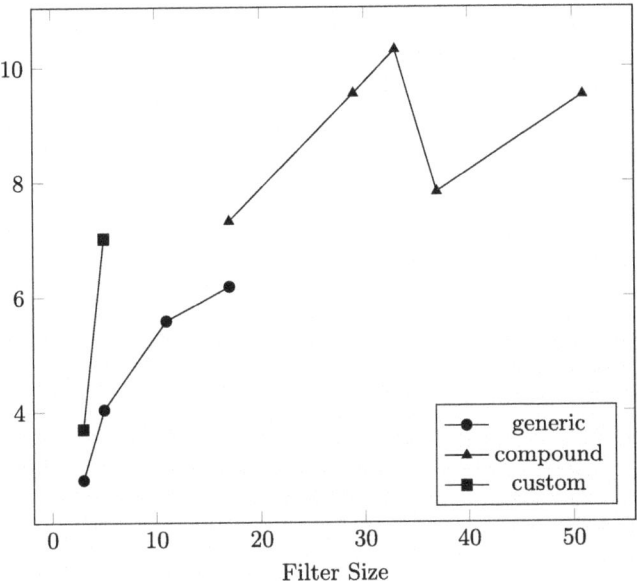

Fig. 1. Speedup of the 2-D Convolution.

An interesting observation happens at filter size 17 as it could be handled by either hardware-specific or compound implementation. The compound variation is significantly faster. It is worth studying this phenomenon closer in hopes of improving the performance of the hardware-specific code and bringing the whole left part of the graph higher.

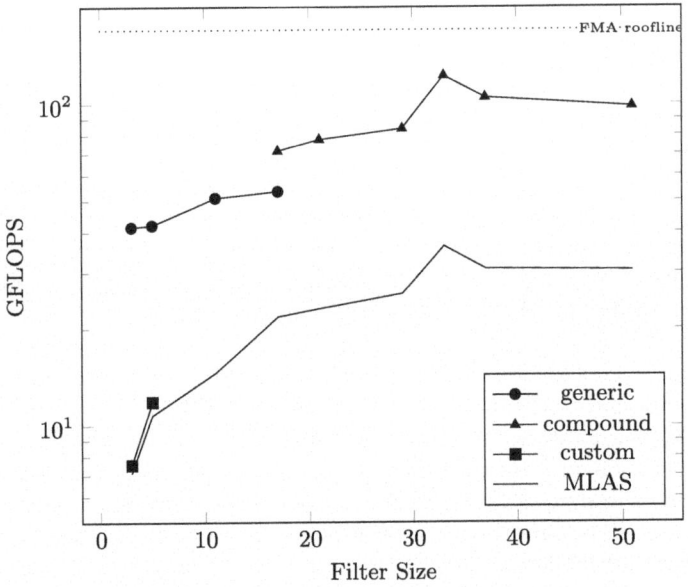

Fig. 2. 2-D Convolution throughput.

The number of arithmetic operations performed by the sliding convolution is the same as the naïve or GEMM-based algorithms. Our observations hint that the speedup comes from better memory access patterns.

We have also measured the arithmetic throughput of different kernels using the Intel Advisor [21]. As the filter size increases, the throughput of the Sliding Window convolution kernels approaches the hardware limits. It is also interesting to see that the filter size misalignment with the hardware vector length results in similar performance patterns for both Sliding Window and GEMM kernels Fig. 2.

3 Conclusion

We have measured the performance and throughput of the Sliding Window convolution kernels. They are more efficient than the commonly used GEMM kernels on the CPU and could even outperform dedicated hardware accelerators. Wider adoption of the Sliding Window sum algorithm could promote AI usage on the low power and low memory devices avoiding the expense of specialized hardware.

All the model compression techniques described earlier are equally applicable to Sliding Window computation. Pruning [9] and distillation [10] reduce the work required for ML inference. Quantization [7] delivers the same benefits of memory and power savings, and better vector performance.

Optimized network architectures tend to use small convolution filters that diminish the advantages of the Sliding Window convolution. In the extreme case of ShuffleNet [32] its pointwise convolutions do not benefit from the new algorithm at all. In general, small filter convolutions are memory bound, equally limiting performance of custom accelerators and CPU solutions. We encourage new research into the network architectures that use fewer layers with larger convolution filters.

In many cases the hardware accelerators can be repurposed for Sliding Window algorithms with various degree of success depending on how specialized the hardware is.

Pipelined nature of the Sliding Window algorithm ensues a straightforward FPGA [31] implementation. Limited on-chip memory and logic resources can be a constraint for implementing large-scale deep learning networks. Combining Sliding Window techniques with optimized network architectures and model compression results in fast and energy efficient solutions.

The algorithms are easily portable to GPU as well. The benefits of streamlined memory access are less pronounced since explicitly controlled on-chip memory hierarchies make GPUs already highly efficient in GEMM computation.

Since the accelerators for matrix multiplication are already present in the current generation of hardware and are likely to stay in future devices, they could improve throughput and performance of many computational tasks beyond GEMM. Thus, it is important to re-formulate our algorithms in terms of the small matrix multiplication, completing the circle. Competition between CPU algorithms and hardware accelerators would lead to advances in both directions, and the most spectacular results are expected at the intersection of the two research fields.

Sliding Window convolution algorithms exhibit excellent performance using commodity hardware. They deliver the benefits of AI to more low-cost and low-power devices.

References

1. Anderson, A., Vasudevan, A., Keane, C., Gregg, D.: Low-memory GEMM-based convolution algorithms for deep neural networks. arXiv preprint arXiv:1709.03395 (2017)
2. Mali gpus (May 2021). https://developer.arm.com/ip-products/graphics-and-multimedia/mali-gpus
3. Bai, J., Lu, F., Zhang, K.: ONNX: open neural network exchange (May 2023). https://onnx.ai/
4. Choi, J., Wang, Z., Venkataramani, S., Chuang, P.I.J., Srinivasan, V., Gopalakrishnan, K.: Pact: parameterized clipping activation for quantized neural networks. arXiv preprint arXiv:1805.06085 (2018)
5. Collobert, R., Bengio, S., Mariéthoz, J.: Torch: a modular machine learning software library. Tech. rep, Idiap (2002)
6. Courbariaux, M., Bengio, Y., David, J.P.: Binaryconnect: training deep neural networks with binary weights during propagations. Adv. Neural Inf. Process. Syst. **28** (2015)
7. Gholami, A., Kim, S., Dong, Z., Yao, Z., Mahoney, M.W., Keutzer, K.: A survey of quantization methods for efficient neural network inference. arXiv preprint arXiv:2103.13630 (2021)
8. Guo, Y., Yao, A., Chen, Y.: Dynamic network surgery for efficient DNNs. Adv. Neural Inf. Process. Syst.**29** (2016)
9. Han, S., Pool, J., Tran, J., Dally, W.: Learning both weights and connections for efficient neural network. Adv. Neural Inf. Process. Syst. **28** (2015)
10. Hinton, G., Vinyals, O., Dean, J.: Distilling the knowledge in a neural network. arXiv preprint arXiv:1503.02531 (2015)
11. Howard, A.G., et al.: MobileNets: efficient convolutional neural networks for mobile vision applications. arXiv preprint arXiv:1704.04861 (2017)
12. Hubara, I., Courbariaux, M., Soudry, D., El-Yaniv, R., Bengio, Y.: Binarized neural networks. Adv. Neural Inf. Process. Syst. **29** (2016)
13. Iandola, F.N., Han, S., Moskewicz, M.W., Ashraf, K., Dally, W.J., Keutzer, K.: SqueezeNet: alexnet-level accuracy with 50x fewer parameters and <0.5 mb model size. arXiv preprint arXiv:1602.07360 (2016)
14. Intel movidius myriad x (May 2021). https://www.intel.com/content/www/us/en/products/details/processors/movidius-vpu/movidius-myriad-x.html
15. Jia, Y., et al.: Caffe: convolutional architecture for fast feature embedding. In: Proceedings of the 22nd ACM International Conference on Multimedia, pp. 675–678 (2014)
16. Jouppi, N.P., et al.: In-datacenter performance analysis of a tensor processing unit. In: Proceedings of the 44th Annual International Symposium on Computer Architecture, pp. 1–12 (2017)
17. Li, Z., Liu, F., Yang, W., Peng, S., Zhou, J.: A survey of convolutional neural networks: analysis, applications, and prospects. IEEE Trans. Neural Netw. Learn. Syst. **33**(12), 6999–7019 (2021)

18. Lin, Y., Han, S., Mao, H., Wang, Y., Dally, W.J.: Deep gradient compression: reducing the communication bandwidth for distributed training. arXiv preprint arXiv:1712.01887 (2017)
19. Nvidia deep learning accelerator (May 2018). https://developer.nvidia.com/nvidia-deep-learning-accelerator
20. Nvidia jetson (May 2021). https://developer.nvidia.com/embedded/jetson-modules
21. O'Leary, K., Gazizov, I., Shinsel, A., Belenov, R., Matveev, Z., Petunin, D.: Intel advisor roofline analysis. Accessed Jul **28**, 2020 (2017)
22. Sandler, M., Howard, A., Zhu, M., Zhmoginov, A., Chen, L.C.: MobileNetv2: inverted residuals and linear bottlenecks. In: Proceedings of the IEEE Conference on Computer Vision and Pattern Recognition, pp. 4510–4520 (2018)
23. Snytsar, R.: Sliding window sum algorithms for deep neural networks. arXiv preprint arXiv:2305.16513 (2023)
24. Snytsar, R., Turakhia, Y.: Parallel approach to sliding window sums. In: Algorithms and Architectures for Parallel Processing: 19th International Conference, ICA3PP 2019, Melbourne, VIC, Australia, December 9–11, 2019, Proceedings, Part II, pp. 19–26. Springer (2020). https://doi.org/10.1007/978-3-030-38961-1_3
25. Tan, M., Le, Q.: EfficientNet: rethinking model scaling for convolutional neural networks. In: International conference on machine learning, pp. 6105–6114. PMLR (2019)
26. Van Zee, F.G., Van De Geijn, R.A.: BLIS: a framework for rapidly instantiating BLAS functionality. ACM Trans. Math. Softw. (TOMS) **41**(3), 1–33 (2015)
27. Vasudevan, A., Anderson, A., Gregg, D.: Parallel multi channel convolution using general matrix multiplication. In: 2017 IEEE 28th International Conference on Application-specific Systems, Architectures And Processors (ASAP), pp. 19–24. IEEE (2017)
28. Wang, E., et al.: Intel math kernel library. high-Performance Computing on the Intel® Xeon Phi^{TM}: How to Fully Exploit MIC Architectures, pp. 167–188 (2014)
29. Wang, H., Ma, C.: An optimization of im2col, an important method of CNNs, based on continuous address access. In: 2021 IEEE International Conference on Consumer Electronics and Computer Engineering (ICCECE), pp. 314–320. IEEE (2021)
30. Wen, W., Wu, C., Wang, Y., Chen, Y., Li, H.: Learning structured sparsity in deep neural networks. Adv. Neural Inf. Process. Syst. **29** (2016)
31. Zynq ultrascale+ mpsoc (May 2021). https://www.xilinx.com/products/silicon-devices/soc/zynq-ultrascale-mpsoc.html
32. Zhang, X., Zhou, X., Lin, M., Sun, J.: Shufflenet: an extremely efficient convolutional neural network for mobile devices. In: Proceedings of the IEEE Conference on Computer Vision and Pattern Recognition, pp. 6848–6856 (2018)
33. Zhou, S., Wu, Y., Ni, Z., Zhou, X., Wen, H., Zou, Y.: DoReFa-Net: training low bitwidth convolutional neural networks with low bitwidth gradients. arXiv preprint arXiv:1606.06160 (2016)

Author Index

A
Abdelrazeq, Anas 3
Albusac, J. A. 11

C
Castillo-Hermosilla, Mariana P. 95
Chapman, Adriane 73
Cross, Miranda 86

D
Dharmaraj, Mallika G. 46

G
Gerding, Enrico 73
Gmez-Portes, C. 11

H
Herrera, V. 11

K
Kemmerling, Marco 3
Khan, Sadia 29

L
Larasati, Retno 117
Lipińska, Maria 63

M
Markus, Antonia 3
Martínez, S. 11

Mid-dleton, Stuart E. 73
Morales, Alfonso 29
Münker, Sven 3

O
Onitiu, Daria 73

P
Padrón, Marcos 3
Plale, Beth 29

S
Schez-Sobrino, S. 11
Schmitt, Robert H. 3
Scott, Lyndsey 105
Snytsar, Roman 125

T
Tayebi-Jazayeri, Hedye 95

V
Vallejo, D. 11

W
Williams, Jennifer 73
Williams, Victoria N. 95

Y
Yazdanpanah, Vahid 73

SPRINGER NATURE

GPSR Compliance

The European Union's (EU) General Product Safety Regulation (GPSR) is a set of rules that requires consumer products to be safe and our obligations to ensure this.

If you have any concerns about our products, you can contact us on ProductSafety@springernature.com

In case Publisher is established outside the EU, the EU authorized representative is:

Springer Nature Customer Service Center GmbH
Europaplatz 3
69115 Heidelberg, Germany

The manufacturer's authorised representative in the EU is Springer Nature Customer Service Centre GmbH, Europaplatz 3, 69115 Heidelberg, Germany. If you have any concerns regarding our products, please contact ProductSafety@springernature.com

Printed and bound by CPI Group (UK) Ltd, Croydon, CR0 4YY

26/03/2026

02078962-0012